NUCLEAR CALIFORNIA
AN INVESTIGATIVE REPORT

EDITED BY
DAVID E. KAPLAN
INTRODUCTION BY
CONGRESSMAN RONALD V. DELLUMS

**GREENPEACE /
CENTER FOR INVESTIGATIVE REPORTING**

SAN FRANCISCO, CALIFORNIA

First printing, April 1982

Library of Congress Cataloging in Publication Data

Main entry under title:

Nuclear California

1. Nuclear facilities — California — Addresses,
essays, lectures.   1. Kaplan, David E., 1955–
TK9024.C2N8   363.1′79   82-1213
ISBN 0-9607166-0-2

Cover and book design by Naomi Schiff.
Typography by Accent & Alphabet.
Printed by Haddon Craftsmen.

TABLE OF CONTENTS

ACKNOWLEDGEMENTS

I'm especially indebted to the staff of the Center for Investigative Reporting for their long months of support on every aspect of this book. Dan Noyes helped nurse the project through its infancy; his efforts were ably carried on by the work of Mark Schapiro and David Weir.

Likewise, thanks go to the staff of Greenpeace's San Francisco office. Without their support and cooperation, this book would not have been possible. In addition, Greenpeace demonstrated its commitment to independent journalism by insuring that the Center maintained editorial control throughout the project.

Special thanks also go to two people who, while quite unrelated, first sparked the idea for this book. By the winter of 1981, Saul Bloom at Greenpeace was already at work on a booklet about nuclear energy in California. Down south, meanwhile, Judy Coburn was exploring the idea of a "Nuclear California" edition of *New West* magazine.

Alec Dubro, Mark Schapiro and David Weir provided invaluable editorial assistance. Saul Bloom, Doug Foster, Tracy Freedman, Dan Noyes and Stan Sesser also carefully went over the manuscript and made numerous helpful suggestions. Dr. John Gofman lent his considerable expertise to checking for technical accuracy. Mike Kepp's administrative skills saved the book from countless delays. David Pingitore contributed much of the research for the maps and the appendix. Thanks to John Nelson and Jon Steiner for fundraising, Brad Bunin and Warren Green for legal advice, Pamela Eichner and Lynn Pedigo for photography, Catharine Norton for copy-editing and proofreading, Naomi Schiff for numerous consultations on design and production and Pam Eichner, again, for her patience.

Several other groups and individuals also deserve thanks: Joe Ward of the California Department of Health Services; Phil Greenberg of Governor Brown's staff; Alletta Belin and the Center for Law in the Public Interest; Dan Hirsch and the Committee to Bridge the Gap; Jim Hanchett of the Nuclear Regulatory Commission; Liza Cohen and Richard Parker of Parker, Dodd and Associates; Herb Gunther and the Public Media Center; and Danny Moses of the Sierra Club.

Finally, I am indebted to those whose kind contributions to this project and to the Center's general fund enabled us to publish *Nuclear California*: W.H. and C.B. Ferry, Greenpeace, Victor and Lorraine Honig, Liberty Hill Foundation, The Seed Fund, The Tides Foundation and Mr. and Mrs. Warren T. Weber.

David E. Kaplan

INTRODUCTION

CONGRESSMAN RONALD V. DELLUMS

As we move into the mid-1980s we face the largest military buildup of the century. Military planners would have us spend more than $3 trillion on "defense" in this decade, and the proponents of nuclear power would have us construct a staggering number of new power plants throughout the country during the next 20 years. We have strayed far from the rosy future painted in the "Atoms for Peace" plan of the Eisenhower years.

In the 1950s the atom was viewed as the salvation of the future — it promised unlimited energy from clean, safe nuclear power plants and a world safe from war because no one would dare push the button first. But these dreams have turned into a nuclear nightmare. The war in Vietnam showed us that some of our military planners could actually think of using "strategic" nuclear weapons, and at home we have faced the very real nightmare of major accidents at our nuclear power plants.

As you will see from the collection of articles in this book, our state has suffered its share of the accidents and growing risks of nuclear technology. The book details how pervasive this technology is in our lives — and how ubiquitous are the problems. It covers the many facets of nuclear technology in California and as such is the most extensive and useful compendium available to introduce the reader to the subject.

Behind the range of issues which these journalists raise lie other dangers to their work and to our society. This collection of articles represents some of the best investigative reporting being done today. Much of what you read here was gleaned by the authors through painstaking research, numerous interviews and considerable effort to uncover facts that many would have preferred to keep hidden from public scrutiny. Fortunately, our country's Constitution protects the press, and we have a long history of public support for the vital role such protections play in the life of a democracy. Unfortunately, we face a growing threat to these protections.

I was a member of the Select Committee on Intelligence, which was charged by Congress in 1975 to investigate the many abuses of both individuals and the press by the CIA's domestic surveillance, and by excesses of the FBI as well, during the Watergate era. Our work resulted in new controls on domestic spying, in the establishment of independent civilian review and in the establishment of separate committees in both the Senate and the House of Representatives to oversee covert activities. These very important protections are now being seriously threatened by actions from the current administration that would renew involvement by the CIA in domestic spying and make it a crime to publish the names of agents in our intelligence services. Thus any journalist who even inadvertently revealed the name of a member of the CIA would be subject to criminal prosecution. These and other proposals currently pending in the Congress would cripple the work of such journalists as those whose articles appear in this book. It is a loss we can ill afford — and a threat that is only the tip of the nuclear-age iceberg. It is no accident that nuclear technology threatens the very foundations of a democratic society. As the 1975 Barton Report by the Nuclear Regulatory Commission makes clear, there is a growing risk of an emerging police state that underlies every aspect of nuclear technology. There is, moreover, a risk to our economic structure, to our health and even to the continuance of life on this planet.

The current administration is telling the American people that we need the neutron bomb, that we need the MX missile and that we should be willing to sacrifice the money to pay for these weapons. There would be only three ways to finance such a mammoth military expenditure: levy new taxes; increase deficit spending; or cut into the non-military side of the budget. We are already seeing the devastating effects of the first wave of these budget cuts. As joblessness and alienation increase, it is not a sight easy to behold. The second alternative, a massive permanent deficit, is not feasible for an administration promising a balanced budget; and the first alternative is just as unpalatable, given government promises of less intrusion into our lives and our pocketbooks.

It is clear, then, that the cost of financing the military budget will be borne increasingly by the poor, who will suffer most as the government withdraws from a commitment to solve human problems. As we continue to experience the inflationary effects of military spending, the middle class, too, will carry a greater burden of these costs.

The 1980s also will see dramatic proof of the devastating role nuclear technology will play in other aspects of our lives. Cancer will be the hallmark of the decade — the deadly result of years of accidental radioactive emissions, overuse of X-rays, power plant accidents, freeway spills and a myriad of other nuclear hazards to public safety.

The costs of such nuclear accidents will threaten to bankrupt the economy. We will have to pay for increased health care, for dismantling of power plants (for which there is currently no viable provision) and for solutions to the overwhelming problems of nuclear waste disposal. Congress and the administration have buried their heads in the sand in avoiding the enormous social, environmental and human costs inherent in nuclear technology. We will pay dearly for these costs if we escape the even greater risk of this decade — that of nuclear war.

We live in a country where such Strangelovian concepts as "unlimited first-strike capabilities" and "limited" nuclear war are matters of serious discussion by our leaders. We can no longer tolerate the fantasies of "war games" planners and their political allies, who have been conditioned to think in terms of acceptable levels of mega-deaths. The current leadership of this country views the world through the narrow confines of an ideological confrontation between the United States and the Soviet Union. It is a sad reality that many of my colleagues in our nation's capital took a picture of the world in the 1950s and threw away the camera; they view the world and its problems primarily in military terms.

The problems of the world of the 1980s, however, have to be solved politically, economically and socially. America will never again single-handedly and unilaterally dominate and control the world. Nuclear superiority or military superiority in the world is a fleeting thing. The Soviet Union has demonstrated the intellectual, technical and scientific capacity to build whatever we build, and this is a fact that we challenge only at our fiscal and physical peril.

My own conviction is that the policy which must guide us in the '80s ought to be the defense of America, not the domination of the world. National security must incorporate the real security that is to be found in building links with the people of the world, not in an outmoded national security state.

If we are to build a new policy, which rests on peace and not on war, we must forge new alliances and build a new awareness of the dangers inherent in our present course. We must take the information made available in books such as this and share it — using the multiplying effect to build a network of solidarity at home and abroad. With a recent poll showing that more than 70 percent of the American people are now convinced that nuclear war is a real possibility, the field is ripe for the massive national organizing that is our only hope.

The courage, dedication and perseverance toward these goals is manifest in the work of *Nuclear California*'s authors. This book is ample proof that the work of individuals can make a true difference to our collective efforts to bring peace to this world.

Southern California Edison

PART ONE

CALIFORNIA'S NUCLEAR NIGHTMARES

Outside of global war, they are perhaps the most sinister threats of the Atomic Age: nuclear terrorism, nuclear accidents and a nuclear police state. The following three articles detail these threats, and though they read like science fiction at times, they are grounded in events that have, unfortunately, already happened. California has already had its first nuclear core meltdown; its cities have already been threatened by nuclear terrorists; and the state's police have already conducted widespread surveillance of nuclear critics. These are parts of Nuclear California that are transforming our notions of security, and they eventually will affect us all.

Photo: A reactor containment building at the San Onofre Nuclear Generating Station

A CITY HELD HOSTAGE

MICHAEL SINGER AND DAVID WEIR
WITH BARBARA NEWMAN

This is a true story. Although much of it has the ring of fiction, of doomsday novels and disaster movies, it is real. If you lived in Los Angeles in 1975, you lived through the nation's first full-scale peacetime nuclear alert.

And that's not even the frightening part . . .

> Dear Mr. Hartley:
> There is a nuclear device with a potential of 20 kilotons concealed on one of your valuable properties electronically controlled in Los Angeles County . . .

It is Tuesday, November 4, 1975. Jean Brady is opening the morning's mail. Her boss, Fred Hartley, chairman of the board of Union Oil, is out of town. It is the routine beginning to what promises to be a routine day. About halfway through the stack, she comes to an air-mail envelope without a return address. Curious. She opens it and takes out a single piece of paper. The letter is very carefully typed on a Tangier-Tippen typewriter.

> . . . You will not call the authorities or use any electronic surveillance equipment. You could trigger the bomb accidentally and it will be very dirty . . .

Jean Brady calls her supervisor, who, in turn, calls Union Oil's chief of security, Paul Schooling. Schooling is already a little jumpy. Union Oil's California property has, in recent months, been damaged by a mysterious bombing, a bombing still unsolved. Schooling reads the threatening letter, but he knows he has no choice. Calls go out to the Los Angeles Police Department and the FBI.

> . . . If you fail to carry out all of these orders we will send a copy of this letter to the news media with instructions to evacuate California, we will then trigger the device, there will be billions of dollars of damage and Los Angeles will be sterile for a long time.
> —*Fision*

Surprisingly, in view of the fact that "Fision" has placed a time limit of six days on the note, the Los Angeles Police Department and the FBI begin squabbling about jurisdiction. Although the FBI has precedence in the case under the Atomic Energy Act, LAPD Chief Ed Davis tells the FBI to "keep the hell out." Then L.A. Sheriff Pete Pitchess, who has been recently assigned to coordinate the county's emergency response plans, finds out about the crisis and jumps in. It takes all of November 4 and most of November 5 for the FBI to flex its federal muscle and take charge of the case.

It is now 3 p.m. on the afternoon of November 5, 30 hours after the letter to Hartley was opened. The FBI, now running the show, plays it by the book — its officers call in officials of the Energy Research and Development Agency (ERDA) to determine whether or not the threat is credible.

ERDA's analysts note that the letter is not a hurried job — it is clear and rational. They note that the sum of money requested, $1 million, is not impossibly large. They note that the instructions for delivering the money (small bills — nothing over $100 — in two suitcases) are realistic. The money would fit. It is also possible, as the letter claims, to build a 20-kiloton bomb and conceal it. It could be transported in a van without being detected. "Fision" says that the bomb can be detonated by remote control, and this, too, is credible — some bombs can be detonated by a simple telephone call from another city. In addition, the fact that the threatener refers to the bomb as "dirty" adds credibility: Crude bombs *are* dirtier (they disperse more radioactivity per kiloton) than military bombs.

Within 50 minutes, ERDA responds to the FBI: Yes, it's a credible threat. There may be a bomb.

Officials at ERDA's round-the-clock Emergency Operations Center also make a few other calls: the White House Operations Center, the National Military Command Center and the Joint Congressional Committee on Atomic Energy. Now officials at the highest levels of government know that ERDA thinks that there is a possibility that Los Angeles will blow up sometime within the next 108 hours.

It is now 6:30 p.m. on November 5. The decision has been made at the FBI to search for the bomb. There are six major Union Oil properties in Southern California: a skyscraper in downtown Los Angeles; two tank farms, in Torrance and San Pedro; one tank farm refinery in Wilmington; and two dockside port facilities in the L.A. harbor. There is not enough equipment and manpower in Los Angeles to complete the search quickly and thoroughly. The FBI asks ERDA for the services of the Nuclear Emergency Search Team (NEST), a little-known government organization designed to cope with just this sort of situation.

At 6:45, ERDA officials call the NEST headquarters in Las Vegas. Ironically, when the order to "scramble" comes, NEST members are already gathered in a conference planning a mock nuclear emergency exercise. Senior operations officer Jack Doyle remembers, "Suddenly, everybody's adrenalin was pretty high."

By 7:30, Doyle and two others are airborne, on their way to Los Angeles. All night long, special planes are dispatched to pick up NEST members and bring them to L.A. Bomb experts are flown in from nuclear weapons laboratories in New Mexico and Northern California. By dawn, 30 operatives have been assembled for the search — and 45 more are on the way.

It is now 8:20 a.m. on November 6. As the official ERDA chronology has it, "ground search commenced." NEST secretly deploys its personnel to the six Union Oil sites. Only top management and security agents at Union Oil are alerted. For all the other workers, it's just another day on the job.

NEST sets up its own emergency command post at Los Alamitos Air Station in Garden Grove, installing four special unlisted telephones for communication with field operatives. All day long, helicopters equipped with sensitive detection pods buzz over Union Oil tank farms and port facilities. Men in business suits stride briskly down the corridors of Union Oil's corporate headquarters carrying black briefcases. Inside those briefcases are miniaturized, supersensitive Geiger counters tuned to wireless earplugs. When a radioactive source is discovered, the NEST operative will hear its "signature" buzzing in his ear.

At 2:30 p.m. a NEST searcher picks up a radioactive "signature" from one of the executive office suites. The source is soon discovered: a souvenir chunk of uranium. At 4 p.m. a searcher at another plant discovers a "hot" railroad car parked on a spur just outside the perimeter of company property. Armed with detectors, NEST operatives gingerly open the railroad car to peer inside. Nothing. The signal was apparently produced by radioactive residue on the floor of the car.

Meanwhile, ERDA is directing other phases of the operation. Naval and LAPD ordnance disposal units are ordered to stand by, although neither of the units has ever had to dismantle a nuclear bomb before. The FBI is developing a plan for trapping the extortionist, but it is clear that the L.A. bureau is unprepared for this big an emergency — one agent is repeatedly called away to deal with bank robberies in his district.

It is now 8:30 p.m. on November 6. All reports from the field are still negative. In Washington, it is near midnight. Sleepy ERDA officials are surprised when one of the agency's top administrators stops by the Emergency Operations Center for advice. What should he tell the press? Everyone agrees that a low profile must be maintained. Even though the threatened 20-kiloton bomb is about the size of the one that destroyed Hiroshima, killing or injuring 150,000 people and leaving a four-and-a-half-square-mile area around ground zero completely burned out; even though L.A.'s famous inversion layer might trap the radioactivity in the basin and create an atomic cloud no one could survive, ERDA officials quickly agree that the danger of panic is worse than the possible consequences of not releasing the information. An official scribbles in the official chronology, "No public release."

As a consequence, during this week in November, the citizens of Los Angeles read in the *Times* that John Robinson is the new USC football coach, that Francisco Franco is fading fast, that Patty Hearst's trial for bank robbery will soon begin; but they do not read that the government is taking very seriously a nuclear extortion plot directed against their city.

NEST continues its covert search. The constant buzzing of helicopters near Long Beach prompts residents to complain to City Hall officials, who demand to know what's going on. Federal officials brief them and urge them to keep the secret. They agree.

It is now 3 p.m. on November 7. All Union Oil sites have been searched. There is nothing more that NEST can do. If the extortionist has been smart enough to shield the bomb with lead or water, then NEST, with all its sophisticated equipment, will probably not find it. If nothing happens in the next three days, the incident will be officially judged a hoax, and it will be up to the FBI to catch the perpetrator. ERDA orders its field teams to wrap up the operation.

The Unlikely Extortionist

Seven months later, a federal grand jury indicted Frank James, a 63-year-old car-leasing salesman with no previous criminal record, for "use of [an] instrument of commerce to threaten to destroy property." During a brief, virtually unreported trial in October 1976, the details of how the FBI caught the man charged with precipitating the first all-out nuclear alert in this nation's history were revealed.

About 30 agents participated in the operation on November 10, 1975 — the day the extortionist had demanded that the money drop be made. Following instructions given in the threat letter, a classified ad was placed in the *Los Angeles Times*: "Fision, we do not want FO [fallout] on L.A." A telephone number, allegedly Union Oil's but in fact one of the FBI's many unlisted numbers, was included.

That evening, FBI agents observed James placing a call from a phone booth at the same time another agent, posing as a Union Oil executive, received the extortionist's instructions for the money drop.

The following day, agents swooped down on James' houseboat in San Pedro. Their thorough search failed to find any physical evidence linking James to the plot. Nevertheless, James was convicted of "threatening to destroy property" — not quite the same thing as extortion — and became the only person ever found guilty in a nuclear threat incident. "As far as the government was concerned, the less said about it the better," the assistant U.S. attorney who prosecuted the case remembers. James' appeals lawyer says his client is "totally inconsistent with the profile [of a nuclear extortionist] developed by the feds." James was nonetheless sent to prison, but he was released within six months after suffering two heart attacks. To this day, he maintains that he was the victim of an "FBI frame-up."

The implications of the case are alarming. If James is guilty, then the question becomes: What drove him to it? Why would a man with no previous criminal record, no grievances against the government or Union Oil and no interest in politics suddenly decide to hold up an entire city for ransom? And how was a man with no special knowledge of nuclear bombs able to write such a convincing note, a note that galvanized the entire government into action?

Look at it the other way. If James is not guilty, if the FBI was participating in a frame-up or merely mistaken, then there's a nuclear extortionist still out there, a terrorist who probably has some sort of nuclear expertise, perhaps waiting for another chance.

But all right. Let's assume that the FBI got the right man. Let's assume that the government was right when it decided not to tell the citizens of Los Angeles about the threatening letter and the possible consequences. Now the question is: Why did the government take it all so seriously? Isn't a nuclear bomb hard to make? Isn't the technology of nuclear explosions beyond the reach of all but the most sophisticated governments?

The answer is no.

Nuclear weapons are easy to make. At least 10,000 physicists, and virtually any well-educated person with the will to learn certain details, are capable of designing — and making — such a weapon.

A Very Simple Bomb

For years it was accepted that the military had a monopoly on nuclear bomb–making. Indeed, it is impossible for a single individual to build a military-style bomb, in part because of the need to convert plutonium or uranium to a "weapons-grade" metal, a process that involves a series of expensive engineering, chemical and metallurgical techniques. Another difficulty involves the construction of the trigger, or implosion system, which consists of a series of high-explosive charges that are placed strategically all around the metal sphere surrounding the nuclear material. They must detonate within microseconds of each other and with equal force and pressure in all directions in order to compress the nuclear material to the point where a nuclear explosion occurs. The complex system is essentially an engineering miracle.

The complexity of these two steps gave many weapons experts confidence that a crude bomb could not be built. However, a four-month investigation conducted for this story reveals that:

• The necessary calculations to build such a bomb are within the reach of second-year physics students. With proper reference materials the calculations can be done in 30 minutes.

• The materials necessary for constructing the bomb are available in a well-stocked hardware store, at a cost of less than $2000. Tools required to thread, cut and shape the metal can be rented.

• One fissionable material, plutonium oxide, requires no processing. The amount needed would fit in a household bucket.

• It is possible to detonate plutonium oxide with high explosives from a *single source,* thus eliminating the need for complex implosion systems.

• One person, working alone, could build a bomb in two easily assembled sections, each light enough to be carried by a person of average strength. The bomb could be taken to a target site where the fissile materials are stored and could then be armed in minutes.

• The total volume of a one-kiloton bomb could be less than 14 cubic feet; the bomb could fit in the trunk of a car (as opposed to the convoy needed to transport 1000 tons of TNT).

• Construction of the bomb would require no highly technical procedures. The few dangerous steps involved in handling high explosives and fissile material can be circumvented with information available to the public.

• Even more frightening, there is a strong possibility that an infinitely simpler nuclear device could be constructed. This device would require no mathematical calculations at all, no special tools, and would be even dirtier than the crude nuclear bomb mentioned above. While we were unable to obtain definite confirmation of the feasibility of this astonishingly simple nuclear device, at least three of the experts we interviewed confirmed that such a device is entirely within the realm of possibility.

The information that we have gathered about the size and nature of these crude nuclear devices is far more specific than we are willing to reveal. For obvious reasons, we do not feel it is appropriate to print the exact plans for these "kitchen-table" bombs. In this matter, you will have to take our word for it — or, more precisely, you'll have to take the experts' word for it. These things can be built by anyone. As one government official commented, "It's so damn crude, it seems impossible. But it gives me nightmares because I'm afraid it would work."

The Single-point Syndrome

Interviews with many weapons experts have led us to conclude that if a terrorist gained access to plutonium oxide he or she could detonate that material in a relatively simple way. Sources stated that plutonium oxide could be detonated from a single high-explosive source, thus eliminating the need to build a complex implosion system. This radically simple method of detonation is known as single- or one-point initiation.

The implications of this formulation are devastating. Plutonium oxide can be stored right now in as many as 18 sites around the country: There are three such sites in California, one in New York and three in Pennsylvania. If obtained by a terrorist, this material could be poured into a prefabricated nuclear bomb in minutes and detonated from a single point. The high explosives needed for single-point initiation can be purchased by almost anyone, almost anywhere in the United States. A crude bomb may not disintegrate a city, but it could take off the top 100 floors of the World Trade Center or pulverize every government building in the immediate area of L.A.'s Civic Center. Its radioactive fallout would leave a deadly substance in the lungs of many suburban dwellers in the vicinity.

The government will not publicly comment on the credibility of any bomb design. However, several knowledgeable sources, including General Mahlon Gates, manager of the Department of Energy's Nevada underground testing program (testing above ground stopped in 1962), and Donald Kerr, head of Los Alamos Laboratory in New Mexico, where prototypal nuclear weapons are designed, conceded in interviews that it might be possible to get a fission explosion from a single-point initiation using plutonium oxide. (Documents released under the Freedom of Information Act indicate that in some cases, single-point tests have *indeed* yielded nuclear explosions.) In fact, Gates oversees a program called single-point safety testing.

In addition, while government officials may not publicly acknowledge the reality of crude bombs, they've got names for them: In nuclear establishment lingo the bomb is a CFE (clandestine fission explosive), IND (improvised nuclear device) or "gadget." It is a weapon that a political terrorist, criminal or any other alienated person can use. In the government lexicon, such people are "nonstate" or "subnational" adversaries.

Government officials who do know all the details relating to the construction of such bombs have misled the public about how easy it is to design and build them. Experts who have tried to confront the government about this problem have been ignored, harassed or muzzled. The government explains that such actions are necessary to protect national security. But we have discovered other reasons for the secrecy:

• The government's ability to discredit any nuclear threat is severely hampered by the fact that so much bomb-grade nuclear material is unaccounted for that it can never again be certain this material is not in the hands of a terrorist or a deranged individual.

• If and when a crude nuclear bomb is built, the government may not be able to detect its location.

• Even if the bomb is found, experts may not be able to disarm it.

- Bomb-grade uranium and plutonium remain inadequately protected because the Nuclear Regulatory Commission (NRC) has not provided sufficient safeguards at nuclear facilities.
- If a crude nuclear bomb is detonated in a major urban area, government analysts predict the following consequences: mass deaths and radioactive contamination and the possible suspension of certain constitutional guarantees.
- The country's vulnerability to nuclear terrorism, however great it is now, will dramatically increase under the "plutonium economy" — the next stage of nuclear power, planned for the 1980s, in which plutonium is the primary source of energy.

The Plutonium Perplex

By 1976, more than 100 threats of violence had been made against American nuclear reactors, including one by the hijackers of a Southern Airways DC-9 on Oak Ridge National Laboratories in Tennessee. There also have been acts of sabotage against nuclear installations, such as the $5 million fire at the Indian Point reactor in New York, started by a disgruntled employee. There remains, too, the threat of sabotage or terrorism on shipments of high-level nuclear waste which regularly travel the nation's rail and trucking lines. The situation abroad is at least as serious. A nuclear plant was seized by guerrillas in Argentina. Large quantities of nuclear materials have been stolen in India. And Israel has now bombed an Iraqi reactor into rubble.

It is the use of a clandestine nuclear bomb, however, that may pose the greatest threat of nuclear terrorism. During the 1970s at least 55 nuclear bomb threats were made against American cities, including one in San Francisco. Four of them, including the one in Los Angeles, triggered high-level responses. One of these was in Orlando, Florida, where a 14-year-old boy eventually admitted to making the threat. Another occurred in Boston, Massachusetts, on April 26, 1974, when officials received this note:

> We are taking this opportunity to inform you that a certain quantity of plutonium has fallen into our hands for reasons we don't feel we have to go into . . .

Government officials have quietly set up an elaborate operation in the DOE called the Emergency Action and Coordination Team (EACT), which exists to counter the threat of nuclear terrorism. Among its components is a threat assessment team to analyze all available information; the highly mobile NEST search teams headquartered in Las Vegas and in the Washington, D.C. area; and a weapons analysis group of bomb designers from the Los Alamos and Lawrence Livermore nuclear research laboratories. The bomb experts are responsible for determining how a clandestine bomb is designed, what its likely yield would be and whether it could be dismantled.

EACT is organized around the progressive stages of a nuclear threat. In the first stage, EACT assesses the credibility of the threat. Is the message grandiose and hurriedly composed, or is it rational and consistent?

If EACT determines that a threat is credible, its next defense is to send out the NEST team. NEST is equipped with extremely sensitive detection equipment that can be carried in briefcases, trucks or helicopters. The devices can pick up signals from any radioactive sources. But, as Jack Doyle, senior operations officer for NEST, says, "You can certainly plan scenarios where this approach doesn't get it." The reason is that anyone can shield nuclear material or a bomb with a sheet of lead, as an extortionist claimed to have done in a 1976 nuclear threat in Spokane, Washington. And as Ted Taylor, formerly one of the top U.S. nuclear bomb designers, said, "It's also easy to shield with water — down in a swimming pool. . . . To say we have instruments to fly around with a high assurance of finding a bomb — that's false."

If a NEST or FBI search does happen to locate a nuclear device, EACT is faced with a whole new set of problems. The team must get close enough to the bomb to analyze what's in it and how it might work. Taylor says a bomb might be impossible to inspect because of "any kind of booby-trapping that would signal a physical approach to the weapon. Or it could be a bomb that requires a code detonator, and if you start manipulating the thing, it's geared to be detonated."

There are two possible outcomes to an attempt to disarm a nuclear bomb, says Donald Kerr of the Los Alamos Laboratory: "You either become confident that you can disarm it, or you begin to figure out how to mitigate the consequences." Unable to disarm or move the bomb, the government would have to detonate it under controlled conditions. Such an event would demand extraordinary measures on a scale unknown in the history of this country.

There is another scenario that leaves the population and the government completely powerless. "We may not get an indication of the thing until we see the first flash," Doyle worries. "The guy may just punch it off and say, 'See what I've done here? Now I want to make you a deal.' "

As a matter of policy, the U.S. government refuses to make concessions to terrorists. That policy is rooted in the belief that once the government gives in to one terrorist's demands, others will seize the opportunity to force similar concessions.

Robert Kupperman, until recently the chief scientist of the U.S. Arms Control and Disarmament Agency, has studied the possibility of a terrorist threat to a major city. "Should the destruction of a major city be considered a real possibility, the government would have little choice but to consider major concessions," Kupperman believes — "concessions that could potentially undermine its ability to govern." And because of what he sees as a dangerous lack of preparedness, Kupperman fears that the government may overreact to the first credible threat — resorting to "repression on a broad scale."

After the Blast

In the aftermath of the accident at Three Mile Island, the government and the media have focused public attention on the problem of making nuclear power plants safe. But there is a deeper, more troubling question underlying the nuclear debate that is not publicly addressed: What kind of world will the nuclear society create?

"The focus on safety issues is incorrect," says NRC Commissioner Victor Gilinsky, "because the true vulnerability of the system is in nuclear safeguards. Safeguards are the thorn in the side of nuclear power."

The reason is simple. To do a conscientious, responsible job of safeguarding bomb-grade nuclear material, government officials feel they might have to assume powers that will fundamentally alter the nature of our society. They will need to wiretap, bug and open the mail — perhaps on an everyday basis — of those whom they suspect might be nuclear terrorists.

The 1975 Barton Report, commissioned by the NRC, envisioned the evolution of a nuclear police state. It predicted the emergence of a special police force empowered to conduct domestic surveillance without a court order, to detain nuclear critics and dissident scientists without filing formal charges and, under certain circumstances, to torture suspected nuclear terrorists.

The Plutonium Economy

As many as 20 or 30 years may pass before we can detect the individual human cells exposed to radiation that have gone "berserk" and caused cancer. In much the same way, we are only now, in the early 1980s, having to confront the social and political fallout of the Manhattan and "Atoms for Peace" projects of the 1940s and 1950s. That legacy, if the nuclear industry has its way, will be the "plutonium economy." All the enriched uranium and plutonium produced to date is insignificant when it is compared to what is planned for the future.

As the end of the first generation of nuclear power plants (those which, in simplified terms, use uranium as fuel and produce plutonium as waste) approaches, we face a second stage, which the industry dubs "plutonium recycle." The concept is simple and logical: "Mine" the plutonium from the waste products of the first-generation plants and "recycle" it as an even more efficient fuel into refurbished old plants or into a new type of nuclear plant — the "breeder reactor."

Plutonium literally "breeds" more plutonium — a regenerating process that might ultimately offer a partial solution to overdependence on other sources of energy, such as oil. But, on the negative side, this means that a vastly increased amount of material would be moving around the country.

The NRC has estimated that as much as 200,000 pounds of special nuclear materials could be produced by 1990, and much of the future plutonium would be in the form of plutonium oxide, giving terrorists or organized criminals the opportunity to intercept the bomb material in transit.

The plans for plutonium recycle and the breeder were blocked by President Gerald Ford, who was concerned about the government's inability to safeguard bomb-grade material. President Jimmy Carter also opposed development of the breeder, but Congress passed authorization of funds — over his objections. Now President Reagan has made it a central part of his national energy policy.

The Beginning of the End

The picture is indeed bleak. We have a bomb that anyone can make; we have a government placing its faith in high-tech devices that are probably inadequate; we have an increasing amount of bomb-grade nuclear material unaccounted for; we have a large number of experts convinced that sometime, somewhere, probably pretty soon, a terrorist bomb will go off and level a large portion of a major American city.

It seems obvious that only a basic reordering of national priorities can prevent one of these grotesque scenarios from becoming a reality. As difficult as it may be, as much economic hardship as it may cause, the fierce momentum of the plutonium economy must be altered and reversed. Already there are a few hopeful signs — the slowdown on the licensing of new nuclear power plants; the possibility that some existing plants may be closed — but halfway, reformist measures are not enough. We have become convinced that the government is trying to regulate the unregulatable, limit the unlimitable. The simple truth is: Nuclear technology is out of control. Any sane governmental policy must be founded on that central fact.

THE MELTDOWN IN L.A.

DOUGLAS FOSTER

After the accident at Metropolitan Edison's Three Mile Island nuclear power plant in March 1979 — the most serious mishap in commercial nuclear development to date — nuclear risks suddenly became front page news. Unknown to the general public, however, a series of similar disasters across the country predated the Harrisburg, Pennsylvania, accident, including a partial core meltdown more than 20 years ago near California's largest city, Los Angeles.

Unlike the Three Mile Island accident, there was no public outcry, no media hoopla, no Congressional scrutiny following the California meltdown. There were simply blithe reassurances of safety from operators who presided over the meltdown and its aftermath, followed by a few cryptic references in the press. The now-defunct Atomic Energy Commission even failed to mention the meltdown in its annual report to Congress. The pervasive attitude of benign neglect was unfortunate; perhaps if the dangers posed by the accident two decades ago had been fully assessed, later mishaps might have been avoided.

The California meltdown occurred at an Atomics International reactor located approximately six miles west of Canoga Park, in Santa Susana, 35 miles north of downtown Los Angeles. The accident could not have been more unexpected, because the plant where it occurred had been celebrated as a model design — safe, efficient and cheap.

Atomics International's "Sodium Reactor Experiment" (SRE) had also been the first nuclear reactor to provide electricity to an entire town. The legendary television newsman Edward R. Murrow paid tribute to the event with an hour-long documentary about the plant's start-up in November 1957. Murrow was clearly awed by the reactor's ability to electrify the town of Moorpark, commenting, "Enrico Fermi once looked at a reactor and said, 'Wouldn't it be wonderful if it could cure the common cold?' Here at Moorpark, a chain reaction that started with him washed dishes and lit a book for a small boy to read."

Despite this auspicious send-off, there were serious problems. Within two years Atomics International's SRE had been ravaged — fuel rods twisted and reactor core damaged. The first operating commercial nuclear power plant also became the first to place the potential dangers in stark relief:

Radioactive gases were released uncontrollably and, as the meltdown proceeded, workers were exposed. Hundreds of barrels of contaminated equipment and radioactive sodium remain as physical reminders of the accident.

More than two decades after the meltdown, it is still not known whether residents of surrounding communities were placed in grave danger at the time. What has been firmly established is that operators of the SRE repeatedly ignored warning signs of serious reactor trouble until high radiation levels forced evacuation of parts of the plant and its eventual shutdown.

Signs of trouble apparently first appeared on July 12, 1959, when operators could not defuse repeated accelerations of power within the reactor. The control rods constructed to rein in these power surges failed dismally, forcing the operators to "scram," or shut down, the reactor by hand several times. Yet, without monitoring the situation or checking for problems, they started the reactor up again — a disastrous move.

When the final power surge occurred, 13 of 43 fuel rods had melted, 81 pieces of uranium fuel lay scattered near the reactor core and at least 10,000 curies of hazardous radioactivity were released into the atmosphere. No one knows how much was really released during the meltdown; the entire inventory of radioactive material at the reactor was more than 200,000 curies, and no inventory of leftover material is available from Atomics International or the federal government.

Wayne Myers, an assistant general manager at AI, later told an interviewer that the accident was "not that much of a hazard, even to local people. By the short period of time it took us to clean up the facility and to recover it, the potential hazard of a major release into the environment was just not there."

In a subsequent study, however, scientists discovered that the SRE's operation had been flawed almost from inception and that the meltdown could have become far worse due to a combination of technical problems and human stupidity. Theos Jardin Thompson, head of the reactor project at the Massachusetts Institute of Technology and a former member of the Atomic Energy Commission, later wrote,

The circumstances which eventually led to this accident began as early as spring 1958, when the first . . . leak occurred. A second leak occurred on Run 8 on November 29, 1958, and problems continued until July 24, 1959. During that time so many difficulties were encountered that, at least in retrospect, it is quite clear that the reactor should have been shut down and the problems solved properly. Continuing to run in the face of a known . . . leak, repeated scrams, equipment failures, rising radioactivity releases and unexplained transient effects is difficult to justify.

Fortunately for the residents of Los Angeles County, high-level fission products from the SRE reactor were apparently captured by the sodium used as a coolant. If the fuel melting had proceeded unchecked, iodine and cesium — among the most dangerous radioactive substances — might have been released, turning a near miss into a public health disaster. Those living nearby faced a potential epidemic of cancer and leukemia, although they were not aware of the risks and were never informed of the potential danger.

But the story of Atomics International and its 1959 meltdown does not end with a simple sigh of relief about a catastrophe mercifully avoided. AI, now a division of Rockwell International, continues to be at the center of an ongoing debate about the nuclear-related health hazards of life in Los Angeles and elsewhere.

The Atomics International office in Canoga Park serves as the head-quarters of Rockwell's Energy Systems Group (ESG), which plays a critical role in the development of nuclear power and nuclear weapons. This huge complex of atomic industries employs 9100 people at plants located around the West. Its past customers include California's two largest utilities, the Bechtel Corporation and the state Air Resources Board.

In Golden, Colorado, Rockwell's ESG operates the federal government's Rocky Flats plant, which manufactures plutonium triggers for nuclear weapons. Rocky Flats is located 16 miles upwind from Denver and has been the site of several serious nuclear accidents, including one fire that ignited 50 pounds of plutonium.

At the Department of Energy's sprawling Hanford plant near Richland, Washington, Rockwell also manages the reprocessing of spent nuclear fuel and the storage of radioactive waste. There, too, accidents have occurred, including the leakage of highly radioactive nuclear waste.

In and around Los Angeles Rockwell operates five sites that handle varying amounts of radioactive materials. By far the most dangerous facility, though, lies in Santa Susana, at a 290-acre "Nuclear Development Field Laboratory." This is where the ill-fated sodium reactor was built. A plethora of nuclear research and development activities are still pursued there, in-cluding construction of testing systems for nuclear power plants, production of test reactor fuel and development of the breeder reactor. The facility operates two research reactors and a plutonium laboratory, and is licensed by the NRC to handle 1500 kilograms of uranium-235 and 3.5 kilograms of plutonium-239.

Another project at the Santa Susana site is "decontamination and decom-missioning," which studies how to dispose of radioactive equipment, includ-ing the damaged sodium reactor. That reactor was permanently shut down in 1964, but it wasn't until 1974 that a decision was made to decommission. This had never been done before, and it turned out to be an enormous operation, costing more than the original price of construction. Today, AI technicians are still decontaminating the building that once housed the reactor and are regularly transporting the highly radioactive parts out of state.

Since 56 percent of the San Fernando Valley's population, approximately 700,000 people, live within 10 miles of the laboratory, there has been growing concern about health hazards associated with the Atomics International operation. Besides the partial meltdown in 1959, AI's field laboratory has been plagued by serious accidents, including a major sodium fire in 1971. That fire exposed 52 employees to potentially hazardous levels of radiation and pointed to the dangers of storing radioactive sodium from the 1959 meltdown on AI property. AI's neighbors fear another sodium fire outbreak, with the attendant possibility of an explosion and a hydrogen fire that would spew massive amounts of radiation into surrounding communities.

Monitoring at AI reveals that levels of radioactivity in the air are five to seven times higher there than in downtown Los Angeles. But since the company's monitoring does not measure the radioactivity by type, it is difficult to assess the health risks. Without establishing the concentrations of particular isotopes, such as plutonium, it is impossible to predict what health risks are faced by San Fernando Valley residents as a result of their proximity to AI's laboratory.

There has been growing pressure from anti-nuclear organizations, such as the Committee to Bridge the Gap, to step up scientific monitoring of AI's sprawling laboratory grounds. Just as Atomics International's laboratory produced the first commercially-operated nuclear power plant — then the first meltdown, the first decommissioning operation, the first hush-hush attempt to gloss over the dangers — it may also yet yield an important new effort to study the myriad health problems associated with nuclear power generation.

THE EMERGING POLICE STATE

DAN NOYES

In August 1977, 50 people were charged with trespassing for their part in a demonstration at the Diablo Canyon nuclear power plant 200 miles northwest of Los Angeles. Two years later, the California Supreme Court dismissed the charges because of police spying and provocation. One of the defendants, an infiltrator from the Santa Barbara County Sheriff's Department, was alleged to have been the only member of the Abalone Alliance to advocate violence at the planned demonstration. Another police spy was also arrested but later disappeared. Both officers were members of a special intelligence group organized specifically for this purpose.

In a sense, this is an ordinary violation of the civil liberties of legal dissenters. But it is more alarming because the intrusion is but one strand in a web of spying, surveillance, disruption and paranoia spun by official and private attempts to protect the nuclear state.

During the last few years, critics of nuclear power — including prominent scientists, environmentalists and elected officials — have been subjected to continued spying and, in some cases, harassment by law enforcement agencies and the nuclear industry. It is a pattern that bears a chilling resemblance to the FBI's COINTELPRO operation against civil rights and anti-war activists in the 1960s and early '70s. The size of this assault on civil liberties in California is revealed by just a few of the cases that have been made public:

- In 1981, the California Attorney General's Office assigned an estimated 10 percent of the agents in its Bureau of Investigation to a special task force on anti-nuclear activists. Attorney General George Deukmejian ordered widespread intelligence gathering on people planning to participate in an upcoming blockade of the Diablo Canyon nuclear power plant.

- In 1980, military and FBI officials detained a crew from television station KQED in San Francisco and seized its film. The reporters had taken pictures of the Concord Naval Weapons Station in an attempt to prove that nuclear weapons are stored there.

- Also in 1980, reporter David Martin was abruptly fired from his job covering the Lawrence Livermore National Laboratory for the *Tri-Valley Herald*. Martin maintains he was forced out of the Livermore newspaper because his investigative reports were critical of the lab's operations.

- In 1979, the U.S. Department of Energy ordered the *Daily Californian*, a newspaper at the University of California, Berkeley, not to publish a letter criticizing the government's handling of *The Progressive* magazine's article on the hydrogen bomb. The government claimed that "restrictive data" on nuclear weapons is vital to the national defense and unprotected by the First Amendment, even if the information is thought up independently by a private citizen.

- In 1978, two members of the Los Angeles Police Department's intelligence division secretly attempted to videotape a city council meeting where opponents of the Sun Desert nuclear plant were testifying. They were discovered and ejected from the room by officials.

- Also in 1978, San Diego Gas & Electric disclosed that it had paid consulting firms approximately $200,000 for information on anti-nuclear groups. Part of the intelligence provided to SDG&E was a "sourcebook" that included a list of anti-nuclear groups, a profile of certain intervenors and details on anti-nuclear issues. One payment went for information on the Clamshell Alliance, located in New Hampshire.

- Starting in 1977, the Los Angeles Police Department infiltrated several anti-nuclear groups in the Alliance for Survival. One police spy, Cheryl Bell, became the president of the Committee on Nuclear Information at California State University in Los Angeles. Bell also monitored public comments by city council members and other officers on nuclear issues and reported these to the LAPD. Another agent, officer Richard Gibbey, helped plan rallies and demonstrations for the group. After being exposed, Gibbey and the LAPD were sued in 1981 by the American Civil Liberties Union for illegal spying.

- In 1971, Pacific Gas & Electric attempted to discredit filmmaker Don Widener, who produced a television documentary highly critical of the nuclear industry's safety record. Widener filed a libel suit against the utility and was awarded $300,000.

- During the 1970s, PG&E paid Research West, an Emeryville, California, detective firm, at least $214,000 to perform duties including "intelligence activities as to terrorist groups." Research West is notorious for attempting, unsuccessfully, to link critics of nuclear power with terrorism. Lawrence Livermore Laboratory has hired this firm as well.

These actions make it clear that critics of nuclear power are regarded as more than mere opponents — they are enemies of the state. In this atmosphere, dissent can be seen as terrorism, and the relationship between a democratic society and nuclear technology grows increasingly uneasy.

The seeds of this relationship were planted in 1954, when the Atomic Energy Act first stated that nuclear energy "for military and *for all other purposes*" was vital to the national security of the United States. Any critic opposed to any aspect of the nuclear establishment became a potential target for surveillance and harassment. Local law enforcement agencies, such as the

LAPD, and corporations with large nuclear investments, such as PG&E, followed the precedent set by the federal government. California utilities and police departments have often photographed anti-nuclear activists at rallies, conferences and large meetings, according to a report prepared for the Center for Law in the Public Interest.

The federal government stepped up its own role in surveillance in 1976, when the Nuclear Regulatory Commission formed an intelligence network specifically to collect and analyze threats to nuclear facilities. It is uncertain whether the NRC has abused its intelligence capability, but it is known that the agency monitors demonstrators and considers them "potential threats."

The FBI apparently has also interfered with the civil liberties of nuclear industry critics, notably in the Karen Silkwood case. Silkwood, a worker in a Kerr-McGee plutonium reprocessing plant in Oklahoma, was investigating the company for a wide variety of health and safety violations, and then died in a highly suspicious automobile accident. In March 1979, Silkwood's family was awarded a $10 million jury verdict against Kerr-McGee. During the trial, it was revealed that the FBI interfered with, and apparently tried to discredit, a Congressional investigation into the case by Representative John Dingell's Subcommittee on Energy and Environment. FBI documents uncovered in 1977 indicate that the agency collected and disseminated derogatory, unsubstantiated information about Silkwood and Congressman Dingell and his staff. One of the FBI documents asserts that criticism of nuclear power in the Silkwood case was "communist inspired."

It is essential that Americans exercising their constitutional rights in pursuit of an alternative energy policy not be deterred by public or private spies. The FBI, the state Attorney General, the LAPD and PG&E certainly have the responsibility to protect nuclear facilities. But they have not yet learned, apparently, to distinguish between genuine threats and legitimate dissent. At issue is perhaps the most nagging problem posed by a nuclear society: Can a healthy democracy exist within a country that is dependent upon atomic power? For anti-nuclear activists, the answer is already too clear. For others, the answer may become painfully real as we move into a future based increasingly on nuclear power and the nuclear surveillance which apparently must accompany it.

U.S. Geological Survey

PART TWO

HIDDEN DANGERS OF THE NUCLEAR STATE

Too many of the problems in Nuclear California are silent ones. We cannot hear the early rumblings beneath the state's nuclear facilities until a massive earthquake strikes. We cannot feel the invisible power of medical X-rays piercing our bodies until their crippling, cumulative effects appear years later. The silence also extends to a secretive, increasingly crowded traffic in radioactive materials across the state's highways, and to the virtually unknown plans to mine uranium near some of California's prime agricultural lands.

The next four articles detail these hidden dangers of the nuclear state: earthquakes, X-rays, transportation and uranium mining. Together, they expose the quiet preparations now being made which may inadvertently guarantee an epidemic of cancer across Nuclear California.

Photo: A collapsed freeway caused by the 1971 San Fernando earthquake

29

LIFE ALONG THE FAULTLINES

GLENN BARLOW

Californians are no strangers to earthquakes. Nearly everyone in the state can recall at least one small tremor and with it that eerie feeling of watching the light fixtures sway and the dishes rattle. Some less fortunate residents have been caught within the massive rumblings of quakes that devastated San Francisco, San Fernando, Long Beach and other areas of the state.

The seismic hazards faced by Californians have worsened, however, in recent decades. During the past 35 years, scores of facilities which handle radioactive materials have been built across the state. Dozens of nuclear reactors, plutonium laboratories and nuclear weapons depots now sit dangerously close to some of the most powerful earthquake faults in North America. Most of these facilities were constructed before fault zones were adequately mapped and understood. They serve as constant reminders that the next powerful earthquake may contaminate parts of California with lethal amounts of radiation.

It is this single factor, seismicity, that makes Nuclear California vastly different from Nuclear Tennessee or Nuclear New York. Approximately 90 percent of all earthquake activity in the continental United States occurs in California and western Nevada.

Nuclear facilities in California were sited and designed before the concept of plate tectonics was generally accepted. According to this theory, the surface of the earth is made up of gigantic crustal plates, which are moving gradually. Tension is released through earthquakes when the plates collide or slip past each other. The floor of the North Pacific Ocean is moving toward Alaska and sliding against the North American plate. This movement has created the San Andreas fault system, which stretches almost the entire length of California.

The maximum earthquake on the San Andreas Fault could measure 8.5 on the Richter magnitude scale. This could release the power and fury equal to the explosion of 1500 Hiroshima-size atom bombs, according to the California Division of Mines and Geology. By comparison, the 1906 quake that almost destroyed San Francisco measured 8.2; the 1971 San Fernando quake measured 6.5. The Richter scale is logarithmic, so an increase of one number means a tenfold increase in the magnitude of the tremors, and a 30-fold increase in the amount of energy released.

Californians have been warned repeatedly by geologists and public officials of impending seismic disaster. A few of the more recent reports point to the increasing danger we all confront:

• The probability of a major earthquake in California is "well in excess" of 50 percent during the next 30 years, according to a 1980 report by the Federal Emergency Management Agency. The report also notes that local, state and federal agencies are not prepared to respond adequately to the emergency created by the quake.

• A report by the Institute of Governmental Studies of the University of California, Berkeley, using historic quake cycles and recent research, concludes that the probability is higher than 50 percent that a great earthquake will strike California during the 1980s and that it will probably affect major urban areas.

• After the devastating volcanic eruptions of Mount St. Helens in Washington in 1980, the President's National Security Council ordered the U.S. Geological Survey to develop scenarios of possible earthquakes that could severely affect California population centers. The USGS report noted that the probability of a major quake is highest in Southern California and is increasing each year.

The $180 Billion Risk

About halfway between San Diego and Los Angeles, right next to Interstate 5 on the Pacific Ocean, sit the three nuclear reactors of San Onofre. It is one of the world's largest nuclear plants, and it sits in one of the world's most dangerous fault zones. Within 100 miles of the plant live 12 million people — half the population of California. Within four miles of the plant is the Newport-Inglewood Fault, which federal geologists call the most hazardous fault in Southern California. In 1933 that fault slipped and generated the Long Beach earthquake, causing massive damage to areas within 40 miles of San Onofre.

What will happen if a major earthquake triggers an accidental meltdown at one of San Onofre's reactors? We don't know. No local, state or federal agency has ever analyzed the consequences of a simultaneous nuclear accident and severe earthquake. No government agency has ever worked out an emergency response plan for such a disaster.

During the last few years, federal and state agencies have busily drafted disaster plans to use when the next powerful earthquake strikes. In the aftermath of the Three Mile Island accident, they also began making separate plans for major nuclear accidents. But no one bothered to combine the two plans in California. Perhaps these government officials think it is too far-fetched. Perhaps they just don't want to cope with the bureaucratic and environmental nightmare of the ultimate peacetime disaster: an earthquake-induced meltdown at a major nuclear power plant. To get an idea of the extent of damage from such an event at San Onofre, consider the following studies:

• A magnitude 7.5 quake on the Newport-Inglewood Fault, even without taking into account the presence of nearby San Onofre, would cause the heaviest damage of any comparable earthquake in California. According to a 1980 report by the Federal Emergency Management Agency, that quake could kill 23,000 people, injure thousands more and cause $70 billion in damages.

• A meltdown at San Onofre, according to a 1981 Nuclear Regulatory Commission report, could cause 130,000 early deaths and 300,000 latent cancers in people living within 50 miles of the accident. More than one million people would each be exposed to more than 25 rems of radiation. (The allowable dose under routine conditions is .025 rems per year.)

• Another study of a San Onofre meltdown, done by the California Office of Emergency Services in 1980, estimates that the accident could contaminate 16,000 square miles, require emergency health care for hundreds of thousands of radiation victims and force the mass evacuation of eight-to-ten million people. Furthermore, the study places an economic value on the apparent loss of Southern California: $180 billion.

Following the Three Mile Island accident, the President's Council on Environmental Quality attempted to determine what effects a massive release of radiation would have on the population surrounding the nuclear plant. The CEQ concluded that if TMI had melted down, the radiation could have produced thyroid tumors in children as far away as 100 miles. The council recommended that children and pregnant women be evacuated immediately following any nuclear accident, as they were in the Harrisburg, Pennsylvania, area during the partial meltdown at TMI.

But Southern California is not Harrisburg. To evacuate children and pregnant women from the Los Angeles basin under any circumstances is difficult enough to envision, but if San Onofre melted down during an earthquake, the task would become utterly impossible. The quake would probably damage freeways, airports and other routes of escape. Just as the mountain ranges around L.A. trap smog particles, they could also trap clouds of radiation. If the mountain passes were blocked by landslides or other earthquake damage, they would also trap the people trying to escape. Routes to the south would be blocked by the intense radiation emitting from San Onofre.

It was 1963 when the NRC first approved construction of the San Onofre Nuclear Generating Station. Despite the obvious seismic hazards at the site, the NRC waited until 1980 before requiring the plant owners to conduct a thorough seismic investigation. Even before then, however, data began coming in that San Onofre could not withstand a powerful quake from the Newport-Inglewood Fault.

San Onofre's owners, Southern California Edison and San Diego Gas & Electric, were startled in 1975 when the NRC temporarily closed the plant because of seismic hazards. After the structure was strengthened, the reactor was reopened. Meanwhile, construction continued on two additional reactors, and the entire plant was designed to withstand an earthquake of magnitude 6.5.

More bad news for the utilities arrived on December 31, 1980, however, when the NRC issued a new report on earthquake hazards at the site. The report confirmed the existence of an entirely new zone of faults lying directly offshore and beneath the reactors and upgraded the maximum possible jolt from the Newport-Inglewood Fault to 7.0 on the Richter scale. One month later, a new USGS report stated that the Newport-Inglewood could generate a quake with a magnitude of 7.5, an ominous departure from earlier estimates. The San Onofre reactors were never designed to withstand a 7.5 earthquake. Within seconds such a quake could destroy the cooling pipes that bring water to the radioactive core, creating a meltdown which could permanently contaminate Southern California.

PG&E's Seismic Sojourn

Although San Onofre provides the most dramatic example of a combined nuclear and seismic disaster, there are numerous other radioactive sites that lie dangerously close to active earthquake faults. In fact, California nuclear plants have a history of delays, shutdowns and modifications — all because of seismic hazards. Take, for example, the case of Pacific Gas & Electric.

Nuclear License Number One in the United States went in 1957 to a test rector named Vallecitos near Pleasanton in Alameda County. PG&E helped General Electric build the Vallecitos Nuclear Center, hoping it would give the utility a lead in the then-emerging field of nuclear power generation. But, in October 1977, the NRC ordered the shutdown of the largest reactor at Vallecitos. The reason: new geologic evidence showing that Vallecitos could experience earthquakes and ground motions similar to those during the 1971 San Fernando quake. The surface of the ground could rupture as much as eight feet directly beneath the reactor.

General Electric still owns that test reactor and three others at Vallecitos, operates a plutonium lab and handles frequent shipments of highly radioactive spent fuel. Despite the threat of a powerful earthquake, GE is appealing to the NRC to reopen the test reactor.

PG&E went on to other projects after completing Vallecitos. In 1958 the utility began developing a site for a major nuclear plant at Bodega Bay in Sonoma County. Attorney David Pesonen, a longtime critic of nuclear power in California, proved to the Atomic Energy Commission (the NRC's predecessor) that the site was next to the nation's deadliest fault, the San Andreas. The AEC recommended against the license and PG&E pulled out.

Pesonen took on PG&E again in 1968, when the utility sought permission to build a plant one mile north of Point Arena in Mendocino County. Once more, earthquake hazards were documented, and PG&E abandoned the site by 1972.

PG&E experienced similar seismic problems with a proposed site at Davenport, north of Santa Cruz. And the utility's first commercial nuclear plant, built at Humboldt Bay near Eureka in 1963, shut down temporarily in 1977 for refueling, and never reopened because of earthquake dangers.

PG&E's final bout with seismic safety may be the Diablo Canyon nuclear plant, located 200 miles northwest of Los Angeles, near San Luis Obispo. The license to construct these two giant reactors was issued by the AEC in 1966, when the nearest major fault was believed to be 20 miles inland. In 1971, however, a pair of oil geologists named Hoskins and Griffiths discovered an offshore fault that passes less than three miles from the reactors. It was named the Hosgri Fault after its discoverers.

The length of the Hosgri Fault has become a point of much debate because geologists agree that the longer a fault is, the greater the potential magnitude of a quake on the fault. Some believe that the Hosgri is part of a lengthy fault zone that branches off from the San Andreas offshore from San Francisco and continues along the coast south to just northwest of Santa Barbara. If the fault zone is indeed that long, it could produce a quake of magnitude 8.0. Federal geologists estimate that this could generate ground motions at Diablo Canyon as much as three times greater than the reactors were designed to withstand. The NRC, however, decided in 1976 to accept the theory that the fault zone is fragmented and shorter, making it capable of only a 7.5 magnitude quake.

The Second Line of Disaster

The big commercial nuclear reactors, like those at San Onofre, might be sufficiently remote from the epicenter of the coming major quake to avoid meltdown. However, there are enough other nuclear facilities sprinkled throughout the state to ensure that one of them will probably be in the zone of heavy damage. In every densely populated region in California, there are medical, industrial and research laboratories which handle large amounts of radioactive materials. The State of California licenses more than 1800 of these facilities, the NRC more than 200. The list includes 18 research reactors at universities and corporations around the state.

Not included in this list are military reactors or the many stockpiles of nuclear weapons in California. The military, on the basis of "national security," refuses to reveal information on the large amounts of radioactive materials it handles in seismically active areas. While the Navy, for example, claims that it uses "state-of-the-art" building standards at its nuclear facilities, there is no way for a civilian to check this — there are no public hearings, no expert testimony and no appeals when the Navy decides to expand its nuclear arsenals. The Navy, in fact, has never done an environmental impact statement on the potential for a nuclear accident at one of its facilities.

All of the Navy bases in California were sited before they began handling large quantities of radioactive materials. Little or no thought was given to the effects of a major earthquake on nuclear weapons or submarine reactors. This is perhaps best illustrated at the Seal Beach nuclear weapons arsenal, located about 25 miles south of downtown Los Angeles, and built directly atop the Newport-Inglewood Fault. If that fault slips again, as it did in the 1933 Long Beach earthquake, hundreds of pounds of weapons-grade plutonium could be dispersed throughout metropolitan L.A.

Other naval facilities present similar hazards. The Concord Naval Weapons Station, 30 miles northeast of San Francisco, stores nuclear weapons near active fault zones in densely populated East Bay suburbs. The North Island Naval Weapons Station stores nuclear bombs on top of the Rose Canyon Fault Zone, within a mile of downtown San Diego. And the Mare Island Naval Shipyard, near San Francisco, which unloads radioactive waste from submarine reactors, sits atop two fault zones that generated the fierce Mare Island Earthquake of 1898.

In the Livermore Valley, 30 miles east of Berkeley, the Lawrence Livermore Laboratory is located above 13 active faults capable of causing ground motions at the lab, according to a draft environmental impact statement. The lab designs nuclear weapons for the Navy and other branches of the military, handles up to 495 pounds of plutonium at one time and operates a small research reactor (see Chapter 13).

There are further dangers offshore. During the 1940s, '50s and '60s, government and corporate radiation labs in California dumped radioactive waste into the Pacific offshore from San Francisco, Santa Barbara, Los Angeles, San Diego and Mendocino. All of these dump sites are located on complex, active earthquake fault zones. More recently, in 1980, the Navy revealed plans to dump old nuclear reactors from submarines and other vessels off the Mendocino coast. This site is possibly the most active offshore fault zone in California; earthquakes occur there frequently, including a powerful magnitude 7.0 in 1980.

Shake and Bake?

The prospect of a massive earthquake splitting apart a nuclear plant is precisely the kind of problem we wish would disappear. Unfortunately, it won't. We have government scientists telling us that a major earthquake will strike California within 10 years; we have a state covered with nuclear facilities that straddle the most powerful fault lines; and we have regulatory agencies that seem incapable not only of responding to such a disaster, but of even examining what damage might result.

California's shaky history makes it increasingly vulnerable to a major nuclear accident. What will it take to convince federal or state agencies to shut down or at least upgrade the nuclear facilities already located near the state's active fault lines? An accident worse than Three Mile Island? Considering the state's seismic history, such a disaster may not be long in coming.

HOTTEST ROADS IN CALIFORNIA

DAVID KAPLAN

Every day, shipments of highly toxic radioactive materials travel across California highways, railways, waters and airspace. Many are routed through the most densely populated areas of the state. Others are deliberately moved through less crowded, but possibly more dangerous, regions. The shipments include heavily guarded nuclear weapons convoys, massive containers of deadly spent nuclear fuel, freshly fabricated fuel rods for experimental reactors and tons of low-level radioactive garbage bound for dumps out of state.

Transportation is the weakest link in the chain of nuclear safeguards. It is during transport that a serious accident is most likely to occur, when the material is most vulnerable to collision, carelessness or terrorism. It took a Freedom of Information Act request filed by Ralph Nader's Critical Mass Energy Project to expose just how fragile the system is. From 1974 through 1979 at least 450 civilian transportation accidents involving radioactive materials were reported nationwide. One-third of these resulted in spills into the environment. Furthermore, the Department of Transportation estimates that about 70 percent of all hazardous transportation accidents go unreported.

More than 2.5 million radioactive packages were shipped in the United States during 1975, according to a study by Battelle Laboratories. This figure is expected to increase dramatically as the number of new reactors and use of radioactive material grows. Most of these shipments involve a wide variety of medical, industrial and research applications. The list of users includes 13 research reactors and more than 2000 facilities across the state. But by far the largest amount, in terms of both quantity and radioactivity, comes from the commercial power industry and the military's nuclear weapons programs.

Spinning Wheels on Spent Fuel

The most dangerous radioactive materials regularly shipped in California, as elsewhere, consist of spent nuclear fuel. These are reactor fuel rods that have become so contaminated with fission products and so broken down that they will not efficiently support a chain reaction. Federal officials classify the stuff as Special Nuclear Material. It's special because the plutonium inside can be chemically extracted and used to make a nuclear bomb; special because at this stage it's toxic enough to poison an entire city; special because it's high-level nuclear waste.

Spent fuel rods are the most lethal of all radioactive materials now generated in the nuclear fuel cycle. A study by the New York City Department of Health, based on Nuclear Regulatory Commission statistics, indicated that a one percent release from a spent-fuel cask in midtown Manhattan during lunch hour would cause up to 10,000 early fatalities and between 200,000 and 1.3 million latent cancer fatalities. Shortly after release of the study New York City effectively banned spent-fuel shipments.

These shipments are scheduled to dramatically increase as early as 1982. The current average of 300 shipments per year will skyrocket to an estimated 9000 trips annually by the year 2004, according to a recent report by the Oak Ridge National Laboratory in Tennessee. Between 1986 and 2004, says the report, 300 million truck miles will have to be driven to a nuclear waste site in southern Nevada (one of three such facilities nationwide), at a cost of some $15 billion.

Prodded by environmental groups and Congress, the NRC in 1980 finally published the current routes used for spent-fuel shipments. Most of the roads bypass major urban areas. Where they don't, the NRC requires shipments to be accompanied by armed guards to ward off potential hijackers or terrorists. A debate still rages, however, over the best places through which to move spent fuel: on interstate freeways that run through crowded urban areas — where the consequences of an accident could be catastrophic — or through smaller back roads that bypass large cities — where the shipments are most vulnerable to hijacking or sabotage.

Federal agencies responsible for regulating these shipments claim the spent fuel is safe from accidents or even terrorist attack due to the containers used. Each spent fuel cask is a massive receptacle packed with 25 tons of lead and stainless steel. The Department of Energy is fond of showing to the press a dramatic film of one of these atomic barrels being crashed into a reinforced wall at high speed. It survives with only minor damage. The casks are also designed to withstand a 30-foot drop and exposure to a 1475-degree fire for 30 minutes.

Real accident conditions may greatly exceed those of the tests, however, according to critics like the Council on Economic Priorities (CEP), a New York–based public interest group. CEP physicist Marvin Resnikoff says that each cask is like "a pressure cooker on wheels" and that the Transportation Dept. is "tempting disaster" by encouraging their widespread use. Many substances commonly shipped on America's highways, he says, including diesel fuel, propane and butane, burn at temperatures greater than 1475 degrees and can produce a fire that burns for far longer than a half-hour.

Critics like Resnikoff also point out that the Nuclear Regulatory Commission, which oversees cask safety, has repeatedly been forced to recall containers because of faulty design, and that none of the casks now used by

the industry has been subjected to the agency's full-scale testing requirements. Only model or computer simulation tests have been conducted.

Atomic Arsenals on the Move

Along with spent fuel, nuclear weapons rank among the most dangerous
substances that regularly traverse California. Each weapon holds at least
several pounds of plutonium-239 or uranium-235. Plutonium is one of the
most potent carcinogens on earth. A single nuclear weapons accident could
cause a radiological cigar-shaped cloud to spread lethal amounts of plutonium
over 70 square miles, according to the Lawrence Livermore Laboratory near
San Francisco. The military's record, furthermore, gives little reason for
optimism: all but four of the 32 nuclear weapons accidents admitted to by the
Pentagon occurred during transportation.

Every day the U.S. government moves hundreds of nuclear weapons. The
completed bombs are first shipped from Pantex, the Department of Energy's
assembly plant near Amarillo, Texas. The DOE sends the bombs directly to
more than 600 military deployment sites around the world, including an
estimated 12 bases in California. Some are delivered overland in unmarked
trucks, trains and mobile homes (yes, mobile homes), but the DOE prefers to
ship by air, using giant C-141 transport planes. The C-141s averaged 130 of
these shipments each year during the early 1970s.

California plays host, however unwittingly, to an estimated 1200 nuclear
weapons. Once in the hands of the military, these bombs are regularly
moved, by truck and sometimes helicopter, between bases for storage and
maintenance. Periodically the weapons are also sent back to Pantex for refur-
bishing. For this the military uses a special fleet of DOE tractor-trailers that
ship only "weapons-grade" nuclear material. Trucks matching published
photos of these heavily armored semis frequent both the Seal Beach Naval
Weapons Station near Long Beach and the Concord Naval Weapons Station
near San Francisco.

According to official reports, at least eight of these specially designed
vehicles are out on America's roads each weekday. They are equipped with
James Bond–like anti-intrusion devices and elaborate communications gear
and are built like safes. Each shipment is followed closely by a van full of
highly trained, heavily armed security personnel, nicknamed the "Plutonium
Police." A carload of peace activists actually followed one of these convoys
from Seal Beach as it ventured into L.A.'s afternoon rush hour, remaining on
its tail through the cities of Santa Ana, Orange, Anaheim and Riverside.

There is also considerable atomic-bomb traffic along the California coast.
Aircraft carriers, such as the USS Coral Sea at Alameda and the USS Constella-
tion at San Diego, each hold about 100 nuclear weapons, according to the
Center for Defense Information, a widely respected research institute in
Washington, D.C. "I used to bring the USS Providence, a guided missile

cruiser, into San Francisco Bay loaded with nuclear weapons," says retired Rear Admiral Gene La Rocque, director of the organization, "and I used to worry during the fall. The fog was so thick we could easily have run into a ship or an anchor." La Rocque says that 70 percent of the Navy's warships carry nuclear weapons, and that, eventually, "one of them is going to scatter plutonium into a harbor."

La Rocque is also concerned about nuclear-powered ships. These are submarines, cruisers and aircraft carriers which have propulsion systems run by one or more reactors. The ships are floating nuclear power plants, and they regularly steam into California's largest harbors. The public, however, has no way of determining what dangers they pose. The Navy's reactor accident records remain classified.

Turning Back the Nuclear Convoys

The prospect of frequent shipments of nuclear waste and plutonium has not sat well with many communities across the country. In the last few years dozens of local and state governments have enacted laws to regulate, restrict or ban radioactive shipments. On town meeting day in 1977, 38 Vermont communities passed restrictions on nuclear cargo. A proposal by the federal government to store nuclear waste near Cleveland prompted 15 suburbs to enact a ring of transport bans and regulations around that city. In California, the counties of Marin and Humboldt have banned all shipments of nuclear waste. Sonoma County and the cities of Oakland, Morro Bay and Pismo Beach each have restrictions over waste transport. And the state now requires that the California Highway Patrol receives a 72-hour advance notice of any spent-fuel shipment.

None of these laws is binding on the military, however. And despite political noises from Washington, D.C., about the "New Federalism" and the importance of local control, the Department of Transportation is set to enforce comprehensive new regulations for radioactive shipments it issued in January 1981. These will effectively preempt all local authority over the movement of high-level nuclear waste on the nation's highways. The action has alarmed many state and local governments and has moved anti-nuclear activists to charge that the DOT caved in to pressure from the Department of Energy and the nuclear industry. New York City filed the first suit challenging the DOT regulations in March 1981. The state of Ohio soon followed with a separate suit. The case is shaping up as yet another major challenge to the federal government's plans to revive the nuclear industry, and observers expect the dispute eventually to reach the U.S. Supreme Court.

Meanwhile, California will continue to be crisscrossed by shipments of some of the most deadly cargo in the world. If you live in a major metropolitan area of the state, chances are you live near a nuclear corridor. Here's a rundown on the state's hottest highways:

Los Angeles Basin: Nuclear weapons are loaded onto barges at the *Seal Beach Naval Weapons Station* near Long Beach and transferred to larger ships off the coast. Seal Beach also ships weapons by truck through Orange and Riverside counties to the Pantex assembly plant in Texas.

Cruisers powered by twin nuclear reactors regularly dock at the *Long Beach Naval Station*. Low-level waste is unloaded from the ships and sent by truck to a government dump in Beatty, Nevada.

Atomics International in Canoga Park handles shipments of highly enriched uranium and spent fuel to and from various Department of Energy sites nationwide. The company, run by Rockwell International, is also decommissioning a reactor on its grounds and is shipping the highly contaminated parts out of state.

South Coast: Southern California Edison and San Diego Gas & Electric, co-owners of the *San Onofre Nuclear Power Plant*, regularly shipped tons of spent fuel through Orange and Los Angeles counties until July 1979, when the NRC ordered them to stop. The shipments, which averaged once a month during 1980, were then routed through less populated areas. They moved south from the plant along Interstate 5, eastward on California 78 through Oceanside, Vista and San Marcos, and then turned north in Escondido, heading towards Riverside County on Interstate 15. It was the beginning of an eight-day journey to a "temporary" storage facility in Morris, Illinois. The shipments may resume again soon if the courts strike down a 1980 Illinois law which restricts the storage of nuclear waste.

Also conveniently along the route is the Navy's *Seal Beach Annex at Fallbrook*, a major nuclear weapons arsenal next to the Marine Corps' base at Camp Pendleton. Special truck convoys guarded by DOE security teams go the distance from Fallbrook to Pantex in Texas.

San Diego Area: Nuclear submarines, cruisers and aircraft carriers regularly bring their reactors to dock at the *San Diego Naval Station*, the *Point Loma* submarine base and the *North Island Naval Air Station* in San Diego Bay. The complex makes regular shipments of low-level waste from the vessels to Beatty, Nevada. Naval personnel also load nuclear weapons onto warships at the *North Island Naval Air Station* and send and receive the DOE truck convoys from Pantex.

General Atomic in La Jolla receives shipments of uranium and fabricates it into fuel rods which are sent to a specially designed commercial reactor in Fort St. Vrain, Colorado.

San Francisco Bay Area: Nuclear reactors aboard submarines, cruisers and carriers regularly plow through the waters of San Francisco Bay to the *Alameda Naval Air Station*. The base sits next to Oakland, one of the world's busiest ports. These and other vessels transfer their nuclear weapons at Alameda and ship them by truck and occasionally by helicopter to the

Concord Naval Weapons Station for storage and maintenance. The helicopter shipments pass over densely populated areas of Alameda and Contra Costa counties. The Concord base also acts as a terminal for weapons from the DOE assembly plant in Texas.

Instead of using Alameda, weapons are also transferred off the coast using barges from Concord. The barges steam to and from the weapons station, a route that takes them under six bridges, through the Carquinez Straits and into San Pablo Bay.

The Mare Island Naval Shipyard near Vallejo is the only base in California that refuels the nuclear reactors aboard Navy subs and cruisers. Spent fuel is removed from the ships and packed onto slow-moving, carefully guarded trains called "Radioactive Specials" by local railroad workers. The waste moves through Contra Costa County on its way to the Idaho Falls National Engineering Laboratory, a huge federal "nuclear reservation" in southeastern Idaho.

Mare Island receives shipments by truck and train of fresh uranium for its refueling operations and sends large quantities of low-level waste to Beatty, Nevada. Also, the shipyard is starting to decommission nuclear subs, which means that tons of highly radioactive material must be shipped out of state.

The Lawrence Livermore National Laboratory, about 40 miles east of San Francisco, sends and receives considerable amounts of uranium and plutonium. The lab constructs working models of nuclear bombs, which are transported to the Nevada Test Site near Las Vegas. The DOE has aroused local concern by routing plutonium through the Livermore Airport in the past. The lab also generates spent fuel from its reactor and ships it by truck every 12 to 18 months to Idaho Falls.

The Port of Oakland has been designated by the NRC as the West Coast shipping terminus for nuclear power plant waste. The city of Oakland, in fact, is the largest city in the country placed on a spent-fuel route by the NRC. Oakland's port now handles spent fuel and other nuclear waste about once a month and moves an average of three radioactive shipments per month. Nearly all are bound for or coming from Europe and the Far East.

At least half of these shipments are made by *General Electric*. GE mostly ships uranium dioxide, a form of nuclear fuel, to Tokyo. While not terribly radioactive, the fuel supplies a crucial link to Japan's nuclear power industry. Other shipments through the port during 1980 included 50 pounds of nuclear waste from the Honolulu Department of Public Health; an 18-ton "irradiator" from Tokyo filled with uranium-238; more than nine tons of contaminated machinery from Tokyo; and 3.5 pounds of radioactive krypton-85 and strontium-90 packed in two wooden crates bound for Europe.

GE regularly ships spent fuel through the port and trucks it to and from its *Vallecitos Nuclear Center* near Pleasanton. Before the NRC shut down the large test reactor at Vallecitos (because of earthquake hazards) the company made spent-fuel shipments every 14 to 16 months. GE still makes frequent shipments of the stuff, however, using NRC-ordered routes that run along Highway 4 and into Highway 160. These roads pass through the fragile areas around Sherman Island, where the highway runs almost 20 feet below sea level in the dike-surounded lands of the San Joaquin Delta. A "worst-case" accident here could contaminate the water supply for 14 million people, because the water is pumped to Southern California, the Central Valley, San Jose and much of the East Bay region.

The Central Valley: This fertile agricultural region is traversed by spent-fuel shipments from Lawrence Livermore, Vallecitos and Mare Island. The NRC routes move from Highway 160 to Highway 12, passing Travis Air Force Base and running east along Interstate 80. Near Vacaville the road branches, with a northern route on Interstate 5 toward the government's nuclear waste facility in Hanford, Washington. There is also a southern route that loops around Sacramento on Highway 113 and then moves out Highway 65 to Roseville. There it joins Interstate 80 and travels into Nevada.

The Navy ships spent fuel from Mare Island by rail. The trains pass right next to downtown Sacramento, through Roseville and over the valley to the Sierra. *The Rancho Seco Nuclear Power Plant*, which currently stores its spent fuel on site, will have to begin shipments out of state around 1986.

Nuclear weapons in the valley are shipped to and from Pantex, in Texas, from *Mather Air Force Base* near Merced and probably the National Guard unit at the *Fresno Air Terminal* and the nearby *Lemoore Naval Air Station*.

Other Areas of the State: Pacific Gas & Electric's *Diablo Canyon Nuclear Power Plant* near San Luis Obispo, should it come on-line, will eventually have to ship its spent fuel to a permanent storage facility. PG&E's *Humboldt Bay Nuclear Power Plant* near Eureka will probably be decommissioned soon, which means the highly radioactive core must be disassembled and shpped, along with the remaining spent fuel, out of state.

Other nuclear weapons bases that ship to and from Pantex include *George Air Force Base* near Barstow and the *Sierra Army Depot* near Herlong, California, about 60 miles northwest of Reno.

Offshore: Finally, there is the more obvious, but less comprehensible threat of nuclear warheads accidentally zeroing in on California's major cities. The Soviets undoubtedly have targeted their ICBMs and long-range bombers on California. According to Admiral La Rocque, at any one time there are Soviet submarines armed with 300 warheads off North American coasts. (In contrast, adds La Rocque, the U.S. bases some 3000 warheads on subs off Soviet coastlines.) Nikita Krushchev once reportedly told Richard Nixon how the Soviets had to destroy a berserk nuclear-tipped missile of theirs after it reached Alaska. Another accidental launching like that could easily become the ultimate nuclear transportation accident.

THE SILENT EPIDEMIC

MICHAEL SINGER

Every day Californians absorb two million tiny bursts of radiation which are not accidents. They are not from nuclear power plants nor are they from weapons storage facilities. They occur in hospitals, clinics and dentists' offices, in the form of X-rays, and may be responsible for up to 90 percent of all the man-made radiation absorbed by Californians each year.

There are two basic dangers from medical X-rays: Radiation to the body cells can cause cancer, and radiation to the genitals of women and men can cause both cancer and genetic damage to unborn children. Just one spinal X-ray exam on a woman or barium enema exam on a man delivers more than one-half the maximum radiation dosage recommended as safe for Americans over an entire year. And the X-ray doses allowed by the United States are the highest in the industrialized world.

The immediate effects of X-rays are invisible; you can't feel or see them. It is over time, however, that the cumulative effects of numerous X-rays have their impact.

Dr. Karl Z. Morgan, former Director of Health Physics at Oak Ridge National Laboratories, and one of the nation's leading experts on the health effects of low-level radiation, describes our cells as storehouses of information and compares X-rays passing through our bodies to "great flashes of lightning." Some cells die quickly from the insult of X-rays and are replaced. Some survive and repair — the lost information in them is recovered. "But occasionally," Dr. Morgan says, "the cell survives and continues to divide, but does so with insufficient or improper instructions. Perhaps it has lost the information which tells it to stop dividing and becomes the precursor of wildly dividing cells which we call cancer."

Genetic damage from exposing the gonads to radiation can be obvious — Down's syndrome or death of the fetus. More often the effects are elusive. "Future generations," as Morgan puts it, "may lack vigor, be susceptible to disease, have poor physiques." These effects are impossible to trace to a specific X-ray exam, but there is scientific evidence such recessive mutations occur.

There is also mounting evidence that women and children are much more vulnerable than men to the effects of X-rays. Pioneering studies by Dr. Alice

Stewart of Oxford University in the 1950s showed that children X-rayed *in utero* had, until the age of 10, a 40 percent higher chance of contracting leukemia than children who were not X-rayed *in utero*. Nonetheless, says Dr. John Villforth, director of the U.S. Bureau of Radiological Health, "300,000 women in the U.S. annually receive X-rays which directly expose the fetus to radiation. Many of these exams are needed, but some are not."

Because there is some emphasis in hospitals on getting women to announce if they are pregnant, the impression is created that non-pregnant women are safe from harm to a potential fetus. Dr. Roland Finston, who teaches doctors at Stanford about radiation safety, says, "There is documentation based on human population studies that there is measurable risk of increasing childhood cancers as a result of not only post-conception radiation but from radiation prior to conception from the beginning of the menstrual cycle."

According to a 1979 report by the California Department of Resources, many X-rays are performed with faulty equipment that does not properly limit the radiation doses received by hospital patients. California, in fact, ranked 44th in the nation in 1979 for the frequency at which it inspected X-ray machines for safety. There are 10 inspectors responsible for the state's 17,800 machines, one-third of which are in constant use in hospitals and clinics. According to one state inspector who requested anonymity, the machines are inspected an average of only "once every 10 to 12 years." Patients, he added, probably receive "twice the dose they need" due to inaccuracies in the collimation mechanism, which restricts the size of the X-ray beam directed at the body.

Although the genitals are most sensitive to radiation exposure from X-rays, one state inspector estimated that only half of the hospitals in the state provide effective shields to protect them from radiation exposure. If this is true, close to 10 percent of the X-rays given daily in California hospitals are illegally administered and may be damaging the reproductive organs of the recipients.

Many of these X-rays are as likely unnecessary as they are illegal. Dr. Herbert Abrams of Harvard University estimates that nearly one-third of all X-rays currently given are "defensive" — growing out of doctors' fear of malpractice suits. The claim requirements of insurance companies help to encourage doctors to take defensive X-rays. An X-ray technician at a San Francisco Bay Area Kaiser hospital explains, "I often find women in the first trimester of their pregnancy having an X-ray of their lumbar spine, which gives tremendous radiation, because their back hurts from a car accident. The doctors know there is going to be a suit so they have to protect themselves."

Dr. Albert Childs, an internist in Berkeley, explained another aspect of the problem. "The outcome of the encounter between a doctor and a patient is not entirely dependent upon what ails the patient. If you go to an internist who has an X-ray machine, there's a good chance you'll get an X-ray."

The economic rationale for X-ray abuse isn't limited to individual doctors, however. As Dr. Morgan points out, "Our medical institutions are in financial difficulties and in many cases X-rays are given to patients to provide needed revenue." Even for the most conscientious doctors there is a real dilemma, adds Dr. Ronald Castellino of Stanford: "We try to encourage the curtailment of X-ray studies, but if you cut out 10 percent of your general case load you lose a certain amount of revenue. You can't get rid of a technician or a machine, however. The expense is still there so we are caught."

The Food and Drug Administration found that between 1964 and 1970 a 20 percent increase occurred in the number of people exposed to X-ray procedures while the population only increased seven percent. In the same period there were marked increases in the number of films exposed and the number of exposures per exam. The government believes that all these trends will continue as they apparently have in California.

The scope of the problem is great — possibly 200 million useless, harmful X-rays are given in this country each year. Dr. Morgan may have put the whole issue in the proper perspective when he said, "A reduction of only one percent in the unnecessary diagnostic exposures in the United States would reduce the population dose more than the elimination of the nuclear power industry until the year 2000."

CONTROVERSY AT BLACK MOUNTAIN

MARK VANDERVELDEN

In the foothills of Black Mountain in the Los Padres National Forest, an unusual group of 60 American Indians from 23 tribes makes up the Redwind Community. Here, some 50 miles east of San Luis Obispo, is an assortment of sun-baked kivas, hogans, sweat lodges and a solar greenhouse. It is a serene place, set among the scrub oak and remnants of ancient Chumash Indian culture.

There are several wells at Redwind used to tap subsurface streams, but the Indians no longer use them. Instead, they haul the community's water 40 miles in from the nearest town because they believe their local wells are contaminated with radiation. The people of Redwind may be California's first recorded victims of uranium mining.

Lomex Corporation, a little-known energy company based in Corpus Christi, Texas, has tried since the early 1970s to get at the uranium near Black Mountain. It has worked closely with the U.S. Forest Service on plans to tap what may now be the only economical source of uranium ore in California. In the process, both Lomex and the Forest Service have incurred the wrath of local officials, anti-nuclear groups, American Indians and the State Assembly Minority Leader Carol Hallett. These critics charge that Lomex plans to use a poorly understood and risky mining technology which may permanently contaminate the local aquifer with radiation.

Unlike most uranium deposits found throughout the West, the small amount of uranium ore residing near Black Mountain is concentrated within ancient streambeds, where it gradually precipitated and collected. This relatively unusual geology lends itself to "in situ" leaching of the uranium, of which Lomex is one of the leading developers.

In situ uranium mining takes advantage of the highly water soluble chemistry of natural uranium. The trick is to drill into an ore body, pump a leaching solvent into the ground and then pump the uranium leachate to the surface, where it can be chemically separated. The process has a number of advantages over open-pit or deep underground mining. Because several of the key mining operations take place in the ground, the problem of disposing of toxic ore tailings is largely eliminated. Uranium tailings have been the cause of serious environmental and public health problems in Colorado and Utah, where more conventional mining techniques have been used. Atmospheric releases of cancer-causing radon gas and heavy metals are also considerably lessened by in situ leaching.

Despite these and other advantages, many people in Redwind and surrounding communities are concerned about the mining operation. Like all advanced nuclear technologies, in situ mining has its drawbacks. Drilling into ore bodies and the use of various leaching solutions could permanently perforate and contaminate local groundwater supplies. Such problems have already occurred in Wyoming, where the Nuclear Regulatory Commission was forced to close that state's first commercial in situ mining operation after excess levels of chlorides, ammonia and uranium were found in test well samples.

"The greatest concern," says Assemblywoman Hallet, "is that Lomex's mining is above the headwaters of the Salinas River, and the potential pollution to the water supply for homeowners as well as agriculture throughout the Salinas Valley is immense."

The contamination of groundwater with leaching chemicals and radiation may have already begun at Black Mountain. When Lomex first embarked on a uranium exploration program in the early 1970s, it drilled more than 100 test holes in the area where it now hopes to expand its drilling. Subsequent testing by government experts showed abnormally high levels of radioactivity in the groundwater. The U.S. Geological Survey, for example, has found that some of the Lomex cores exceed state and federal danger levels for alpha radiation.

California health officials still view these radiation levels as "within compliance" because Redwinds water contains relatively low amounts of the radium isotopes which the state uses to measure radiation. But for the people of Redwind and a growing number of citizens in San Luis Obispo and Monterey counties, these are ominous figures. Says Toby Buffalo, an aide to the Redwood community, "The state should be testing for other elements that emit radiation. We just don't know what other substances are present here."

Lomex, meanwhile, has kept a low profile on the problems of Redwind, preferring instead to work through the U.S. Forest Service. The company, in fact, was unavailable for comment on any aspect of this story, despite repeated attempts to reach a spokesperson. "Often they don't even call *me* back," says Keith Guenther, the Santa Maria district ranger handling the project.

Guenther has been under fire from groups opposed to the mining, who have accused the Forest Service of excluding local participation in the project's evaluation. "I've been gravely concerned about the total shutting out of local authorities, from the Board of Supervisors to individuals," says Carol Hallett. "It's something that surprises me as far as the Forest Service is concerned."

Toby Buffalo of Redwind charges that the Forest Service has repeatedly "withheld information, excluded interested parties from key decisions and stopped local jurisdictions" from having a say in the final decision on Lomex's plans. "We find this totally unacceptable," says Buffalo, who warns of a lawsuit unless Guenther disqualifies himself.

Guenther, however, denies that anyone has been shut out. "The county has been involved all along the way," he says, "and I've met with representatives of Redwind numerous times during the last two years. They all have complete access to the data."

Opponents of the mining assert that public pressure finally forced the federal government to agree to an environmental impact statement, the first on a uranium exploration program in the nation's history. The results of the draft EIS should be in during early 1982, but any actual mining may be far in the future. The Lomex Corporation faces a long and treacherous political fight before it can tap the uranium ore near Black Mountain.

Lomex is not alone in its interest in California's uranium. Exxon, Tenneco, Utah International, Tennessee Valley Authority, Rocky Mountain Energy Company and a dozen other energy conglomerates have cornered the bulk of the best mining claims in the state, with Exxon holding the largest and most promising in southern Kern County. But these companies, like Lomex, have as yet been unable to economically extract the ore.

Years of exploration have proven that the state's uranium reserves are not concentrated in a single area, but are spread in remote, environmentally sensitive areas such as the Los Padres National Forest. The economies of scale which make uranium mining profitable in Colorado, New Mexico, Texas and Utah are simply absent, as are the milling facilities needed to refine the ore.

The prospects for uranium mining in California once looked brighter. During the late 1940s, "yellowcake fever" had caught on and was spreading like a fast wind throughout the western United States. In search of yellowcake, or concentrated uranium, would-be fortune hunters took Geiger counters in hand and headed for the deserts and hills. Bolstered by federal price supports, exploration bonuses and other financial incentives, this band of enterprising prospectors located somewhere between 200 to 300 sites nationwide. It was the beginning of a multibillion-dollar industry that would fuel America's programs in nuclear weapons and power, while creating serious environmental problems for the workers and communities which helped it grow.

In California there were a number of small but active commercial mining claims by the mid-1950s. Petersen Mountain in Lassen County, the Coso range in Inyo County, the Miracle Hot Springs area near the Kern River and parts of the Mojave Desert all produced small amounts of commercial grade ores.

The most productive uranium mine in California was the Juniper Mine, situated near the Sonora Pass in Tuolomne County. From the time uranium deposits were first discovered in 1956, the Juniper Mine produced an estimated $1.6 million worth of uranium. Although mining at Juniper is unlikely to resume, Utah International still exercises an active claim.

By the early 1960s, interest in California's uranium reserves began to wane. Richer, more accessible deposits were being found in New Mexico, Texas and Wyoming. National uranium production peaked in 1960 and continued to decline, along with prospecting, through the 1960s. By 1966, U.S. production had dropped to 10,000 tons, and California's contribution was reduced to just a few hundred pounds of ore.

The market remained relatively stable until the mid-1970s, when an international cartel of uranium producers and suppliers began to press the price of yellowcake dramatically upward. Within two years, California reserves which were uneconomical to mine at $8.50 per pound began to look much more attractive at $45 to $60 per pound. The uranium industry took a hard second look at California.

Then the bottom fell out. The demise of the commercial nuclear reactor market in this country caused the domestic uranium market to crash. Of the 42 million pounds of uranium produced in the United States during 1980, only 20 million pounds entered the fuel cycle, with the remainder going into inventory. Even the military was sitting on a seven-year surplus of yellowcake. Prices plummeted from $50 per pound to $23 per pound. Many industry analysts now predict a further slump and that only mills with firm long-term delivery contracts will be able to operate profitably. A number of mills have already closed their doors, with many others working at only a fraction of their capacity.

During the first half of 1981, 25 uranium mines in the Four Corners region of the Southwest were forced to close, throwing thousands out of work. The world's largest open-pit uranium mine, operated by the Anaconda Corporation, has also been closed indefinitely. All this comes at a time when foreign producers are ready to aggressively enter the international market. Australia, South Africa and Canada all possess large, high-grade uranium reserves. These nations do not have large domestic demands for their production and will be jockeying with U.S. suppliers for the shrinking share of the international market.

The uranium companies are now busy lobbying Congress for new laws which would revive the ailing nuclear industry. The corporations are seeking a ban on uranium imports and an easing of licensing requirements for commercial reactors. But even these drastic measures might not be enough to help the industry in California. Were it not for the persistent efforts of the Lomex Corporation, uranium mining in the Golden State might already be a dead issue. And judging by the poor track record of Lomex, it may yet become one end of the nuclear fuel cycle about which Californians need not worry.

THE NUCLEAR FUEL "CYCLE"

MINING
Uranium ore is dug from underground or strip-mined.

MILLING
The ore is crushed to a fine powder and turned into uranium oxide, or yellowcake.

CONVERSION
The yellowcake is combined with fluorine to produce uranium hexa-fluoride gas.

ENRICHMENT
The gas yields a smaller amount of uranium enriched enough to fuel a reactor or make a bomb.

FUEL ASSEMBLY
The enriched uranium is returned to oxide and fabricated into fuel rods for use in reactors.

CIVILIAN REACTORS
Utilities run more than 70 reactors to produce electricity while dozens more are used in research and industry.

MILITARY REACTORS
The federal government runs nearly 200 reactors at test stations and aboard Navy vessels.

MILITARY REACTORS
The U.S. maintains several reactors to produce plutonium and tritium for nuclear weapons.

REPROCESSING
Utilities have been unable to reprocess spent fuel for use in reactors due to economic problems and the risks of weapons proliferation.

REPROCESSING
Unburned uranium and plutonium are recovered from military spent fuel and used to manufacture nuclear weapons.

WEAPONS ASSEMBLY
Nuclear bomb parts are made nationwide and assembled in Pantex, Texas.

WASTE STORAGE
Contaminated clothing and other radioactive garbage are stored as low-level waste. Spent fuel and other high-level waste is now stored "temporarily" at special federal dumps and commercial reactor sites.

DEPLOYMENT
The weapons are transported by truck, train and plane to an estimated 600 sites worldwide.

Information for this chart was compiled by the Center for Investigative Reporting with assistance from Jim Harding and Howard Morland. Design by Naomi Schiff.

51

PART THREE

CALIFORNIA'S NUCLEAR ARSENAL

California is a land surrounded by the impulse of nuclear war: to the east, the American military regularly explodes atomic bombs to be used against the Soviet Union; to the west, Soviet submarines offshore aim ballistic missiles at our largest cities. Inland, too, we find the makings of a mighty nuclear war machine: a dozen military bases armed with the latest in atomic weaponry.

No one doubts that we need a strong defense. We can no longer deceive ourselves, however, about the real costs of preparing to fight World War III. The next four articles explore in chilling detail the impact of California's nuclear arsenal, from the leaking reactors of the nuclear Navy to the hundreds of nuclear weapons regularly handled, and mishandled, throughout the state. Also included is the story of California's atomic veterans — Americans who have already lived through a kind of nuclear war — and, finally, a depiction of the grim realities we face in the event of a nuclear attack on the state.

Photo: "Operation Crossroads" at Bikini Atoll in the mid-Pacific, July 1946

WHERE THE BOMBS ARE

DAVID KAPLAN

In a desolate area of this nation's nuclear test site in Nevada, federal officials built a life-size model of a California town during April 1981. The town's sole purpose was to act as a stage for a "Broken Arrow" — a major nuclear weapons accident. Members of the military, the California Highway Patrol and the state Office of Emergency Services played along on a script taken from real life: a helicopter crashes while carrying at least three nuclear weapons. There is a fire and one of the weapons is ruptured and leaking plutonium.

The accident was the second in a series of carefully planned simulations called NUWAX, the military's code name for Nuclear Weapons Accident Exercise. The seven-day, multimillion-dollar event was arranged by the Pentagon's Defense Nuclear Agency. By the end of the exercise, the handful of Californians involved became the first state officials in the country to jointly respond with the military to a nuclear accident.

NUWAX is the Pentagon's belated attempt to manage what may have already become an unmanageable problem: the nation's nuclear arsenal. For the last 10 years public attention has zeroed in on the problems of the commercial nuclear industry. Uranium mining, reactor safety, waste disposal — these are the problems we hear on the evening news. But most of this country's use of nuclear power has gone unscrutinized. Most of it escapes the wrath of the anti-nuclear movement. This is because most of it belongs to the U.S. military.

Environmentalists worry about toxic chemicals in their neighborhoods, unaware that there may be hydrogen bombs stored nearby. Nationwide concern erupts over nuclear plants like Three Mile Island, yet no one questions the safety of similar reactors aboard warships that sail regularly into our largest harbors. The facts are, however, that twice as many nuclear reactors are run by the military as by utility companies, that American communities are routinely exposed by movements of nuclear weapons, and that 99 percent of the volume of high-level radioactive waste in this country is the legacy of 35 years of atomic weapons programs.

For most of us the U.S. nuclear arsenal conjures up visions of giant missile silos located in the expanse of the Great Basin, of remote Strategic Air Command bases spread along the Great Plains and of a nuclear-powered navy far away at sea. But much of our nuclear arsenal is stored and handled in or near our largest metropolitan areas. A six-month investigation by the Center for Investigative Reporting disclosed that the Defense Department's nuclear weapons and reactors sit alongside every major city in California. It is a large-scale, secretive, virtually self-regulated use of radioactive materials — and it raises serious questions not only of nuclear safety, but of nuclear accountability as well. Consider:

• There are as many as 19 Navy vessels powered by 29 nuclear reactors whose home ports are in California harbors. Radioactive spills and contamination are known to have occurred, yet the Navy's nuclear accident records are classified.

• The military stores an estimated 1200 nuclear weapons on at least 12 bases throughout California. More than half of these are located in or near major urban areas, and some are sitting on active earthquake faults.

• The Navy transports nuclear weapons by truck, barge and helicopter through heavily populated areas of metropolitan San Francisco, Los Angeles and other major cities.

• U.S. military installations have in many cases failed to coordinate emergency plans with local communities and refuse even to acknowledge the presence of nuclear weapons in our cities.

Broken Arrows and Bent Spears

The NUWAX exercise at the Nevada test site was a simulated Broken Arrow, one of the more severe accidents that can befall a nuclear weapon, according to the Pentagon's classification system. The least worrisome of the accidents is a Dull Sword, or a "nuclear weapon minor incident." A "nuclear weapon significant incident" is called a Bent Spear. A Broken Arrow is an official "nuclear weapon accident." And the worst possible blunder, a "nuclear weapon war risk accident," is appropriately called a "Nucflash" (pronounced "nuke flash").

After years of secrecy justified on national security grounds, the Pentagon in 1976 released a list of 27 Broken Arrows, including several that spread radiation over large areas. According to these records, the last Broken Arrow occurred in January 1968, when a B-52 bomber with four hydrogen bombs crashed near Thule, Greenland. In 1980, prodded by continued requests from the public, the Pentagon released a more detailed description of the accidents and mentioned an additional group of "fewer than 10" that are still classified. Oddly enough, all of the accidents listed are from the Air Force. Is it possible that the Army and Navy, which each handle thousands of atomic weapons, have never had so much as a Dull Sword? Apparently not. The Navy recently acknowledged the existence of classified documents dated from March 1973 to March 1978 entitled "Summary of Navy Nuclear Weapons Accidents and Incidents." The documents are hundreds of pages long.

How many accidents have really occurred? The Stockholm International Peace Research Institute, a prestigious think tank funded by the Swedish government, estimates that there were 125 U.S. nuclear weapons accidents, major and minor, between 1945 and 1976, or about one every three months.

As far as the public knows, no nuclear weapon has ever accidentally exploded. The Pentagon maintains that the accidental explosion of a nuclear weapon is so remote a possibility as to be negligible, but federal officials haven't always been so sure. A number of serious Broken Arrows in the early 1950s spurred the Atomic Energy Commission (AEC) to conduct a series of experiments at its Nevada test site between 1955 and 1958. Nuclear weapons were dropped from aircraft, exploded with dynamite, set afire and crushed in speeding vehicles. The results were a little unsettling: the AEC released data on 19 tests "which resulted in a measurable yield."

The risk of plutonium contamination from a wrecked bomb is more worrisome to some public officials than the less likely prospect of a nuclear blast. A typical atomic weapon contains at least four and a half pounds of plutonium; inhaling one-millionth of a gram of it can cause lung cancer. It has a radioactive half-life of about 24,400 years, so you don't want it spread around. That's exactly what could happen, though, if the high explosive were to accidentally detonate because of a crash, fire or mishandling. A radiological cigar-shaped cloud spreading out over 70 square miles could be caused by a single nuclear weapons accident, according to Lawrence Livermore National Laboratory near San Francisco.

To minimize the chance of that happening, the newest weapons have been equipped with "insensitive" high explosives, instead of the more volatile TNT, to trigger the nuclear reaction. This is a wise precaution, since a single accident could have catastrophic consequences. "If enough of the plutonium is vaporized from a weapons explosion," warns radiation expert Dr. John Gofman, a former associate director at Livermore, "and if that then gets lifted up in the atmosphere and goes toward a populated place like the Bay Area, you could guarantee that between 30 percent and 50 percent of the people would be condemned to getting lung cancer."

A nuclear weapons accident is most likely to occur during transportation, when the device is most vulnerable to collision, carelessness or terrorism. Every day the U.S. government moves hundreds of nuclear weapons by air, land and sea. Warships full of the bombs plow through California's largest harbors. Navy bases in crowded urban areas transport them by barge, truck and helicopter. And a busy, highly secretive traffic in nuclear weapons exists between the Department of Energy's assembly plant in Pantex, Texas, and bases in California. These bombs are moved by C-141 transport planes and by heavily guarded convoys of armored trucks.

Are You Living Next Door to an H-Bomb?

It is the official policy of the U.S. Department of Defense to "neither confirm nor deny the presence of nuclear weapons on any ship, station or aircraft." "We don't want to make information available for targeting by the enemy," says Lt. Col. Dale Keller, a spokesman for the Defense Nuclear Agency. "Why make it any easier for them?" The logic of this argument is compelling, but does it correspond to reality? At a congressional hearing in 1973, Major General Edward B. Giller, the Air Force's assistant general manager for national security, was asked if the Russians knew where our nuclear weapons were stored. "Most of our stockpile sites have characteristic fence lines, lighting and communication facilities," he answered. "I would imagine they know where most of them are."

If the Russians do know — and their satellites can detect details as fine as weapons bunkers — then who is all this secrecy aimed at? Congressional sources say confidentially that the problem is now largely one of public relations — that military bases are more concerned about alarming local communities than about tipping off foreign agents.

Using publicly available sources, William Arkin, an independent defense analyst formerly with the Center for Defense Information, estimates that California plays host (however unwittingly) to about 1200 nuclear weapons. That is a lot of firepower, and whether or not we sleep more safely because of it, a lot of it is located very close to home.

Some of the most powerful bombs and missiles in the world are undoubtedly stored at three Strategic Air Command bases: Mather Air Force Base near Sacramento; Castle Air Force Base near Merced; and March Air Force Base near Riverside. The Air Force is also believed to maintain tactical, or short-range, nuclear air-to-air missiles at three air defense bases: George Air Force Base near Barstow; Castle Air Force Base; and the National Guard unit at the Fresno Air Terminal. (Fresno's main airport is divided between commercial and military use.)

The Army stores nuclear missile warheads and projectiles for artillery at the Sierra Army Depot near Herlong, California, about 60 miles northwest of Reno. The Marines guard a major nuclear storage site at Camp Pendleton in Fallbrook. (Fallbrook is also the alleged headquarters for a Ku Klux Klan guerrilla training group.)

The largest handler of nuclear weapons in California is the Navy. In San Diego, home port of about 29 percent of the entire U.S. fleet, the Navy's nuclear arsenal at the North Island Naval Air Station is situated a mile from downtown. The facility lies even closer to San Diego International Airport.

The Navy also reportedly stores nuclear weapons right in the middle of the San Francisco Bay Area, at the Alameda Naval Air Station, near downtown Oakland and next to one of the world's busiest ports. Nuclear depth charges are probably stored in the Navy's anti-submarine warfare center at Moffett Field, a few miles outside of San Jose, and at Lemoore Naval Air Station in the Central Valley. But the major arsenal for Northern California is at the Concord Naval Weapons Station, about 30 miles northeast of San Francisco. Concord's claim to fame is a 1944 accident that blew off 3.5 million pounds of explosives, killing 323 people, leveling the town of Port Chicago and damaging 12 other cities.

Most of the bases were sited years ago in relatively remote areas, but the suburbs have grown so quickly that homes now sit literally across the street from nuclear weapons bunkers. At Concord, aqueducts that provide drinking water to more than a million East Bay residents run within a mile of the nuclear storage area. Many of the bases sit on or near active earthquake faults, where they might not be allowed to be built today. Should these communities remain less informed than the Russians about the potential hazards in their midst? Nowhere is this question more relevant than at the Seal Beach Naval Weapons Station near Long Beach.

The Seal Beach Naval Weapons Station is the Navy's largest nuclear arsenal in Southern California. It sprawls over some 5000 acres, about 25 miles south of Los Angeles. When the base was founded in 1944 the area around it was mostly farmland and marsh. But L.A. has grown, and the Seal Beach area is now so congested that 10 airports now lie within a 20-mile radius. There are five layers of air traffic around the base below 3000 feet, including the approaches of at least 257,000 aircraft annually. Since the base opened, three planes have crashed into it, including one that left a wide, flaming path of jet fuel not far from the nuclear weapons stockpile.

Further complicating matters, the Navy built the base directly on the Newport-Inglewood Fault, which the U.S. Geological Survey rates as having the greatest potential for damage in the L.A. area. These dangers prompted the General Accounting Office, in a little-noticed 1975 report, to suggest that the Navy consider moving the base. The Navy, however, claims that the Seal Beach Naval Weapons Station presents little hazard to the public and says it has no plans to move.

Nightmare Island

The headquarters of the California Office of Emergency Services (OES) sits on the outskirts of Sacramento in a comfortable single-story building, where the state's emergency planners work. At the heart of OES is the California Warning Center, a large room packed with 20 telephones, six teletypewriters and a sophisticated radio communications system with immediate access to any place in the country. An OES staffer is on duty 24 hours in the center, waiting for a disaster — an earthquake, a flood, a toxic spill, a nuclear accident.

Attached to the wall of one of the larger offices is a huge map encompassing the northern quarter of San Francisco Bay and parts of four Bay Area counties. The map centers on the Mare Island Naval Shipyard, just across the Napa River from Vallejo, a city of 78,000 people. Carefully drawn around the shipyard are concentric circles, indicating evacuation zones. They stretch into Napa, Solano and Contra Costa counties. Next to the map an OES planner works on coordinating the first joint emergency plan for the counties in the event of a nuclear disaster at Mare Island.

Anti-nuclear activists refer to the shipyard as "Nightmare Island." It is one of two places on the West Coast where vessels in the nuclear Navy are refueled. Mare Island and the Puget Sound Naval Shipyard in Bremerton, Washington, handle some 40 percent of all atomic refuelings in the Navy. The pressurized light-water reactor used in these ships served as the model for the commercial nuclear power industry.

The nuclear submarine is an amazing machine. Although less than one-tenth the size of a typical commercial plant, the single reactor aboard most subs enables them to travel on a single refueling for as long as 13 years. They can stay continuously submerged for months. The subs equipped with Poseidon and Trident ballistic missiles remain the backbone of a powerful nuclear deterrence force. A single Poseidon sub, submerged and practically undetectable, carries 16 fully armed missiles, each with as many as 14 independently targeted warheads. There is enough atomic firepower on one submarine to incinerate 160 Soviet cities.

The fleet must be maintained, however, and the peacetime costs of that might be paid in more than dollars. Sub reactor operators have been turning up in the offices of local anti-nuclear groups and even at anti-nuclear demonstrations, seeking information on the health effects of radiation. The National Association of Atomic Veterans, an organization of radiation victims, is repeatedly contacted by shipyard workers worried about overexposure. A former radiological control officer, stationed at the base for almost two years, claims that nuclear accidents at Mare Island were common — and commonly went unreported:

> Workers got contaminated frequently, often twice a month, including myself. We had one spill of about 80 to 120 gallons of primary coolant water. It was a crew member screw-up, very similar to Three Mile Island. It took an hour and a half before the alarm went off, two and a half hours before the shipyard responded. Of the three radiation monitors, only one worked to measure contamination. Every other ship that came in there had some kind of accident like that — containment problems with spills or leaks.

During congressional hearings on nuclear safety after the 1979 accident at Three Mile Island, Admiral Hyman Rickover was asked about the Navy's safety record. Rickover, the "father" of the nuclear Navy, testified that "in the 26 years [of naval reactor programs] there has never been an accident involving a naval reactor, nor has there been any release of radioactivity that has had a significant effect on the environment."

Rickover's critics, including retired Rear Admiral Gene La Rocque, director of the Center for Defense Information, point to a series of problems that paint a different picture:

• Civilian workers have filed at least 58 injury claims due to radiation exposure from the nine shipyards doing nuclear work for the Navy. Claims from Mare Island alone account for 25 of these.

• In 1979, a controversial cancer study by the *Boston Globe* of the Portsmouth Naval Shipyard in Maine — a base similar to Mare Island — revealed that nuclear submarine workers were dying of leukemia at a rate more than 400 percent higher than the national average.

• According to a leaked naval inspection report on Portsmouth, the shipyard is a radioactive mess, with a dangerously high rate of radiation incidents, unnecessary worker exposure, inaccurate radiation surveys and poor record-keeping. Fully 80 percent of the technicians taking a written exam on radiological control failed to achieve a passing score.

What worries state officials most about the Mare Island Naval Shipyard is what they call a "worst case accident." The scenario goes something like this: to refuel a nuclear submarine, the reactor must be shut off and put through a "cooling down" period of between one and three months. The sub is then hauled into dry dock and positioned so that a large crane can be lowered into the reactor core. The crane lifts out dozens of individual "fuel modules," which at this stage are intensely radioactive — full of plutonium, strontium, cesium and other poisonous elements. The crane carefully lowers each module into a massive steel cask lined with lead. If all goes well — and the Navy claims it has performed 166 refuelings without a hitch — fresh fuel modules are lowered into the core and the reactor is sealed. But if anything should go wrong — a crash, a machine failure, an earthquake — the module could rupture and spew radioactive particles into San Francisco Bay and a dozen communities downwind.

Nuclear submarines are constantly in dry dock at Mare Island for overhaul as well as refueling. The ships are propped up four and a half feet above ground by blocks that might not hold during a strong earthquake, according to one engineer who worked at the base. State officials aren't really sure what that would do, because accident studies of naval reactors are classified.

The Navy claims that the spent reactor fuel taken out of submarines is not "stored" at Mare Island. It is, however, kept on base for between one and four weeks before it is shipped out of state. As many as six refuelings are performed annually, so high-level radioactive waste may be present for up to 24 weeks of the year. Government personnel load the heavy casks of spent fuel onto special Southern Pacific trains bound for a DOE facility in Idaho Falls, Idaho. Unlike commercial shipments of spent fuel, however, these trains bear no signs or placards warning of their contents.

The Navy's Nuclear Garbage Dump

The refueling station at Mare Island is but one facet of a naval nuclear force that fairly bristles in California's three major harbors. The Point Loma submarine base in San Diego can service as many as 21 nuclear subs at one time. The larger San Diego Naval Station is home port to at least three huge nuclear-powered cruisers, each with two reactors aboard. Long Beach Naval Station and Alameda Naval Air Station also service nuclear-powered vessels. Alameda is home port to the nuclear-powered aircraft carrier *USS Enterprise.* It is powered by eight reactors, each able to provide an estimated 250 megawatts of power. That's roughly equivalent to a floating Diablo Canyon nuclear power plant with the difference that the *Enterprise* is operational.

The general public was rudely awakened to the Navy's environmental impact last year when headlines screamed that barrels of radioactive waste were leaking in heavily fished waters off the Farallon Islands near San Francisco. That, however, was but the tip of a slowly dissolving iceberg. For instance:

• Nuclear-powered ships have accidentally spilled reactor coolant into San Diego Bay on several occasions during the last two years: 30 gallons in July 1980, 13 gallons in September 1979, another "small leak" in May 1979. The Navy claims there were no health dangers at any time, but when local reporters probed for details a spokesman told them, "Anything having to do with the operation of nuclear equipment is classified information."

• As recently as 1977 the Navy had radiation-monitoring devices at the Hunters Point Naval Shipyard, a seemingly innocuous base located in San Francisco. Officials could not explain their presence 10 years after the closing of the radiological lab.

• Contaminated clothing, machinery and other radioactive garbage is regularly trucked from Mare Island, San Diego and Long Beach to a government dump in Nevada. State records show that the Navy made 24 such shipments in 1977. The routes are classified, but it would be hard to get out of these cities without moving through densely populated areas.

• Nuclear ships and subs routinely dispose of low-level waste while at sea. The Navy, in fact, dumped millions of gallons of it into San Francisco Bay and other coastal waters throughout the Sixties. While most of the waste is now disposed of at least 12 miles offshore, some 25,000 gallons are still poured into U.S. harbors annually, according to official accounts. The public is assured, however, that the stuff is harmless. As one Navy recruiter in San Diego said, "Listen, reactor cooling water is so safe you can drink it."

• The Navy is still actively considering how to sink entire subs — reactors included — onto ocean floors as a means of permanent waste disposal. One proposed site is 200 miles off the coast of Mendocino County.

This, then, is the nuclear Navy: nationwide, 145 floating reactors on subs, cruisers and aircraft carriers, and another nine at training centers in Idaho, New York and Connecticut. That is more than twice the number of commercial nuclear power plants now in operation. Barring any unforeseen progress in arms limitation talks, the nuclear Navy is going to grow. Congress has authorized construction of 36 new nuclear subs and two additional aircraft carriers. These vessels will have to be maintained, refueled, eventually decommissioned and possibly even mothballed near major coastal cities. And California's harbors will inevitably remain among the Navy's favorite home ports.

Making California Safe for Civilians

The only critical look at the military's nuclear safety problems by a government agency comes from the General Accounting Office, the investigative arm of Congress. A 1979 GAO report took the Defense Department to task for failing to develop some way to interact with local officials in the event of a nuclear accident. The military could "deal with a few key state or local officials on a classified basis," the report suggested. The GAO also recommended setting up joint emergency plans in all jurisdictions with the "potential" for nuclear weapons storage. This would leave intact the Pentagon's "neither confirm nor deny" policy. It's been more than two years since the GAO report came out, but, says Sterling Leibenguth, the project leader and author of the report, "In terms of coordinating emergency plans nothing has changed. The Defense Department is still the least prepared of any federal agency we examined to cope with an off-site nuclear accident. It's a paradox, because on base they're the best prepared."

Some California communities are growing impatient with the military's uncommunicative posture. "The Navy undoubtedly has nuclear weapons at their ammunition depot," says Coronado Mayor Pat Callahan about North Island. "We haven't been told directly, but we're positive they have." Callahan is concerned because Coronado, a city of 19,000, is wedged between the arsenal at North Island and the huge naval base at San Diego. Weapons are frequently transported through the city's streets. Coronado officials were particularly irked at the Navy in early 1980 for not having informed them of a planned 25 percent increase in weapons storage at North Island. Callahan and the city council invited the Navy to attend a council meeting, but the Navy didn't show. The city hoped to obtain an agreement — and finally did — that its police and fire departments would be informed when the Navy moves weapons through Coronado.

"We've been talking to them about a joint emergency plan," says Callahan, "but we really haven't gotten very far because of the reluctance of the military to admit that there are nuclear weapons in the first place and then to examine how an accident would affect the city."

Coronado's experience is typical of the response local communities have been getting. Concerned over reports of the presence of nuclear weapons at the Concord Naval Weapons Station, the Contra Costa County Board of Supervisors commissioned an investigation into the transport and storage of toxic and radioactive materials. "None of us here knows for certain what is stored there," says Supervisor Nancy Fahden. "The Navy refuses to tell us. If they do have nuclear material then I would like to see the federal government come up with the additional money to help us devise an appropriate emergency plan."

Other groups in California and elsewhere are launching a legal assault on the Department of Defense to make local nuclear facilities more responsive:

- In Los Angeles attorney Leonard Weinglass recently consented to represent the Seal Beach Nuclear Action Group and to try to force an official public disclosure for the first time that nuclear weapons are stored at a particular location, the Seal Beach Naval Weapons Station.
- In Hawaii environmental groups filed suit to stop the Navy from storing nuclear weapons at 48 newly completed bunkers. They argue that the Navy failed to file an environmental impact statement on the site, located a mile from a major runway approach to Honolulu International Airport. The groups won on appeal last July, obliging the Navy to publicly assess the "hypothetical" environmental impact of storing nuclear weapons. However, the Navy finally won on appeal to the U.S. Supreme Court, and retained its privilege of nuclear secrecy.

The State of California is still trying to get a handle on the problem. In 1979 the Brown administration put together the first comprehensive survey of radioactive materials in California, but officials readily concede they still know little about the military's nuclear operations. Legislative squabbling has added to the confusion. In the wake of Three Mile Island the state passed legislation requiring the Office of Emergency Services to make plans for areas around nuclear power plants. But a bill by former assemblyman Floyd Mori to extend this to nuclear research and military facilities was killed in committee in July 1980.

So the essential question remains unresolved: Should the military's nukes be subject to the same public scrutiny and accountability we impose on private industry? The NUWAX experiment in Nevada may be a sign of a more cooperative attitude in the Pentagon, if it is not simply a grandiose public relations stunt. It would be unrealistic, certainly, to expect a state like California — where defense spending amounts to some $18 billion annually, making it the largest single industry in the state — to take unilateral action. Ultimately the issue will have to be decided by Congress, and right now there is little enthusiasm for such a debate on Capitol Hill.

The structural problem here (unless one believes that the military is immune to Murphy's Law) is that it's hard to keep this nation's 30,000 nuclear weapons and an entire nuclear fleet readied for war and not have something go wrong. Until both the United States and the Soviet Union get serious about arms control this problem will only get worse. The United States is currently going into what may become the biggest arms buildup in history: government reactors that haven't operated since the early Sixties must be turned back on to produce enough plutonium for the new Trident, Cruise and MX missile systems. The buildup will only intensify the paradox we already confront. The more we add to our nuclear arsenal, the more Bent Spears and Broken Arrows there are likely to be; the more emphasis we put on national security, the less secure our neighborhoods become — and the greater, finally, is our chance of dying by our own Dull Swords.

Addendum: The Pentagon's Trail of Broken Arrows

The Pentagon admits to 27 Broken Arrows, or nuclear weapons accidents, but independent observers put the figure closer to 125. Here's a small sampling:

B-52 Crashes in North Carolina: Perhaps the closest call this country has ever had occurred in 1961, when a B-52 crashed near Goldsboro. The plane had to jettison two hydrogen bombs, each with the explosive power of 24 million tons of TNT — 800 times the Hiroshima bomb, or 12 World War IIs in one blast. Dr. Ralph Lapp, former head of the nuclear physics branch of the Office of Naval Research, wrote soon after the accident that five of the six safeties had been set off by the crash. A single switch may have saved much of North Carolina from obliteration.

H-Bombs Drop on Spanish Village: In January 1966, a B-52 collided with its refueling tanker and dropped three hydrogen bombs on the fishing village of Palomares, Spain, and a fourth into the Mediterranean. The TNT in two of the bombs exploded on impact, spreading plutonium over 640 acres. The United States destroyed local crops, plowed up tons of radioactive soil and shipped it in 4827 steel drums to a waste dump in South Carolina.

Greenland Air Crash Scatters Plutonium: A B-52 with four H-bombs aboard crashed into a frozen bay near Thule, Greenland. The plane and bombs apparently disintegrated, burning a radioactive hole 300 feet wide and 2200 feet long. The Air Force shipped 1.7 million gallons of contaminated ice and snow to the United States for waste disposal.

Missile Ruptures in San Diego: On February 10, 1970, a guided missile cracked while wedged in a storage magazine aboard the aircraft carrier *Bon Homme Richard*. The carrier was docked at North Island Naval Air Station, one mile from downtown San Diego. A Navy spokesman said the missile was not believed to be nuclear armed at the time, yet 200 crew members were evacuated, while others prepared to take the ship out to sea. Demolition experts removed the 600-pound missile.

B-52 Crashes in Northern California: On March 14, 1961, a B-52 with two nuclear weapons on board ran out of gas and crash-landed near Yuba City, California. The Air Force maintains that the high explosive in the bombs did not detonate and there was no contamination.

Crash Landing Near San Francisco: On August 5, 1950—within minutes of the fifth anniversary of the atomic blast on Hiroshima—a B-29 bomber crashed near a Strategic Air Command base about 40 miles from San Francisco. The Pentagon says that an atomic bomb on board was merely a dummy, but firemen present at the scene claim there was radioactive contamination. The blast injured 60 and killed 19, including General Robert F. Travis, base commander.

Titan Missile Explodes in Arkansas: On September 19, 1980, a Titan ICBM exploded in its silo, blasting a nuclear warhead into the countryside. Air Force personnel searched for hours before finding it relatively intact. Local residents complained for weeks afterward of flu-like symptoms they believe were caused by toxic fumes released by the missile.

REHEARSALS FOR THE HOLOCAUST

DAN NOYES

The *USS George Eastman* and its crew approached San Francisco's Golden Gate, returning home after three months' duty at Operation Redwing in the western Pacific. It was August 16, 1956. The United States government had just concluded a series of 17 nuclear explosions at Redwing — one of the dirtiest dispersals of radioactivity ever recorded. One blast in the series had been the first air drop of a thermonuclear bomb; several tests had been in the megaton range — hundreds of times larger than the bomb dropped on Hiroshima. Official radiation readings listed 378 sailors at Redwing as overexposed.

The crew of the *Eastman* was glad to be coming home, away from the heat, hard work and boredom of Eniwetok and Bikini. But as they reached San Francisco Bay their homecoming seemed more like an episode from the "Twilight Zone." Before them sat a tug completely draped in canvas, its crew members wearing what one eyewitness later called, "white, space-like suits." The tug was prepared for something the *Eastman*'s crew wasn't — radioactive contamination.

The caution was soon justified, to the horror of the *Eastman*'s crew and their waiting families. When Captain Ross Pennington attempted to disembark at San Francisco's Hunters Point naval shipyard, he was forced to turn back. His shoes were too "hot" from walking on the deck of his radioactive ship.

The *Eastman*'s radioactive return was not unusual. The ship's atomic homecomings also occurred after Operation Castle in 1954 and Wigwam in 1955. Hundreds of other ships came home to California ports from the United States' atmospheric nuclear testing in the Pacific between 1946 and 1962. After numerous tests at Bikini, Eniwetok, Johnston and Christmas islands and other locations, radioactive ships routinely returned to California ports to be decontaminated, studied and sometimes sunk in coastal waters. Radioactive equipment was lost or discarded off the coast during these voyages.

At Operation Crossroads in 1946, 123 of the 200 ships present were considered "radiologically suspect." Several "guinea pig ships" were positioned close to the explosions to determine damage and radioactivity from the bombs. These were later returned to the San Francisco and Mare Island naval shipyards in San Francisco Bay to learn methods of decontamination. The now-defunct U.S. Naval Radiological Defense Laboratory at Hunters Point and the Radiological Defense School at Treasure Island were established in San Francisco to conduct research. One guinea pig ship, the cruiser *USS Salt Lake City*, was returned after Crossroads to San Pedro in 1947 and then used for target practice and sunk off the coast between San Diego and Long Beach because it was "too radioactive" for salvage.

Another target ship, the aircraft carrier *Independence*, was anchored in San Francisco Bay in June 1947 and officially labeled a "negligible" radiation danger. Three years later *The New York Times* reported that the *Independence* was still hot and tied up at Hunters Point, "a floating laboratory for research into the phenomenon of radioactivity." The *Times* report went on to say that "plain, old-fashioned 'elbow grease' is still the best and perhaps only way to decontaminate" a radioactive ship. Decontamination, a Navy officer was quoted as saying, was really just "moving the contamination from one place to another." The article concluded that the San Francisco radiological labs experts "feel that as a menace to military forces, [radiation] has been exaggerated . . . troops and ships could quickly move through the bombed area."

Some of the approximately 42,000 men assembled in the Pacific for Crossroads in 1946 later reported both immediate and long-term health effects from their radiation exposure there. Many received an official "joke" at the time: a certificate of induction as a "Brother Pig" into the "Grand Council, Exclusive Order of Guinea Pigs." But the laughter ended long ago. Now no one knows the full human or environmental cost of these tests. No studies were conducted on the impact of the radioactive ships which returned to California waters, nor were the crews or shipyard workers examined for the effects of contamination. The true story of these tests has too often been buried and forgotten, like the *USS Salt Lake City*, in a radioactive grave.

What is known about atomic testing shows the leading role California has played in its continuing story. Consider these facts:

• Two major cities, San Diego and Los Angeles, received significant increases in radiation levels after atomic tests in adjacent areas. San Diego notched higher levels in 1955 when radiation was monitored after Operation Wigwam, one of two blasts that occurred within 500 miles off the coast. Los Angeles recorded higher readings in 1958 after a series of 20 explosions were detonated in Nevada during a two-week period.

• Radioactive fish were repeatedly found in samples taken from California and Japanese fishing vessels during the Pacific testing in the 1950s.

• Radiation from underground testing in Nevada has vented into the atmosphere and posed a danger to California on several occasions. Underground tests in Nevada have continued at the rate of one a month during the first four months of 1981. In September 1980 radiation from one test was monitored 10 miles from the California state line. Officials were unable to determine if the radioactive gas drifted into the state.

• There are persistent fears — discounted by U.S. officials — that underground tests in Nevada could trigger an earthquake in California.

The Ivory Bomb Tower

From the preparations for the first atomic test in 1945, California assumed a pivotal role in these rehearsals for nuclear battle. Dr. J. Robert Oppenheimer conducted his early research into the development of an atomic bomb at the University of California in Berkeley. In 1943, he moved to New Mexico and set up the university's Los Alamos Laboratory, which helped complete the Manhattan Project. The university's contract with the Manhattan Engineering District of the War Department began the secret research in the New Mexico mountains which led ultimately to the atomic bombings of Hiroshima and Nagasaki.

The end of World War II saw the beginnings of the Cold War with Russia. The U.S. military put a priority on nuclear weapons — a priority that included massive atmospheric and oceanic testing. The University of California, through Los Alamos, the Radiation Laboratory in Berkeley, and, later, the Lawrence Livermore Laboratory, became the recipient of millions of dollars in government contracts. Its mission: develop the nuclear know-how to achieve atomic superiority, know-how which would be tested in more than 600 detonations conducted since the end of the war.

The University of California's Scripps Institution of Oceanography at La Jolla also played a critical early role in the Navy's atomic testing program, starting at Operation Crossroads in 1946. When testing began in Micronesia, the military wanted to measure more than the effects of the atomic explosion on the naval vessels; it also wanted to determine the environmental impact at Bikini. A team of 1000 biologists, geologists, oceanographers and technicians, including many Scripps scientists, was assembled for the study. Roger Revelle, a Navy officer and later director of Scripps, was given charge of the "most complicated laboratory experiment ever undertaken," according to an official Scripps account. He subsequently termed the results of the study "the clearest, most detailed pictures of an atoll and its flora and fauna that we possess." This scientific acuity would become a bitter irony for the men who served at Crossroads and felt that they and their families suffered from its radiation. No studies were ever done on how the bomb affected them.

Scripps Institution's tremendous growth after World War II was a direct outgrowth of the military's increasing interest in the oceans. In 1946 the director of Scripps foretold the coming boom: "The war showed how oceanography could be applied." For atomic testing, the Navy needed to know how to use nuclear weapons on and under the oceans, how to defend against their use by the Russians, and how to measure and reduce the impact of radiation from the tests on sea life, commercial fishing and the surrounding environment. Scripps benefited from the military contracts to build up its equipment and fleet, and used radioactive water resulting from tests to study ocean currents, sediment movement and the effects of radiation on sea life.

The relationship between Scripps and the military was also useful for deflecting criticism of the A-tests. In 1957, under growing pressure to end the tests and eliminate the contamination of the planet from radioactive fallout, Scripps director Revelle stressed to Congress that only one-half of one percent of fish caught by the Japanese fishing fleet after tests in 1954 were radioactive. Japanese scientists saw it differently. They emphasized that this amounted to 480 metric tons of contaminated fish that had to be discarded as "not fit for human consumption."

Over the years, Scripps has continued its involvement in various studies with nuclear applications. Project Hydra in 1963 studied methods to detect possible underwater nuclear testing by the Russians. Funding in 1971 went for construction of a floating man-made "island" that might carry nuclear power plants. Several years later Scripps joined other institutions in a lengthy study to investigate the disposal of high-level radioactive waste on the ocean floor. Scripps' present director William Nierenberg is a veteran of the Manhattan Project and a special adviser to the government on a possible future nuclear weapons system: the MX missile.

Bombs Away

By 1949 the military developed a new need for atomic testing. Officials decided to pursue the production of a hydrogen bomb many times more powerful than the atomic bombs used on Hiroshima and Nagasaki. Again, scientists at Los Alamos and Lawrence Livermore were the principals behind the world's first thermonuclear detonation at Eniwetok in November 1952.

The landing of U.S. troops in Korea in 1950 brought a sudden urgency to the tests. The Defense Department needed a site more accessible and less costly than the remote Pacific atolls. An area near Las Vegas was finally selected because of its proximity to Los Alamos, low population density and ease of control by the government. The first Nevada tests were begun quickly in January 1951. California now had atomic testing to the east and to the west.

Along with the military's need to test atomic weapons, the Cold War continued the strict secrecy concerning the tests that had shrouded the first explosion at Alamogordo, New Mexico, during World War II. The public knew little about them and this kept protests to a minimum. National security meant that nuclear testing had to be kept secret from the American people as well as the Russians and Chinese.

Secrecy allowed the U.S. to explode the first deep underwater atomic bomb 500 miles off the San Diego coast in 1955 with only a flurry of public alarm. The 30-kiloton blast at Operation Wigwam was the Navy's attempt to learn how to use and defend against nuclear depth charges. California played a particularly central role in the test. Initial planning began in 1953 at the Navy Electronics Laboratory in San Diego and quickly moved up the coast under tight security to Scripps. There, under an academic exterior, the Navy secretly planned the experiments for Wigwam.

The assemblage of 30 ships and 6500 men was so tightly controlled that a film prepared for emergency release in case of "adverse public reaction" was not shown until 1979 — 24 years later. Secrecy also became the flip side of public relations. The film was prepared so that "certain of the more dramatic aspects [of Wigwam] are deemphasized." The secrecy for Wigwam worked so well that the *San Francisco Chronicle* ran a May 17, 1955, article downplaying the radioactive effects of Wigwam — next to an article contending that 80,000 hereditary changes would occur in children from previous atomic tests.

Secrecy was also useful when fallout from tests in Nevada hit Los Angeles in 1958. Ten days after U.S. officials told the public the radiation readings were "harmless," an official of the National Advisory Committee on Radiation concluded in a secret meeting that "if you ever let these numbers [for radiation] get out to the public, you have had it."

The political climate in California during the 1950s could not have been better for the tests. California Senator Thomas Kuchel announced right before the Wigwam test that it would be conducted in a "sea desert" and that all fish delivered to the Pacific Coast following the test would be checked for radioactivity. Secret Wigwam memos released recently under the Freedom of Information Act show that the government planned to conduct only a brief monitoring of samples of fish delivered to California for commercial sale and that there was uncertainty about the amount of contamination fish might absorb. A press release prepared weeks before the test announced that no radioactivity had been found in fish samples.

California Governor Goodwin Knight sent the manager for the Nevada tests a letter of congratulation in 1955 for the recent completion of a test series. He applauded the tests for bringing "to our state a new appreciation of the potentiality of nuclear weapons." A small photo of an atomic blast was sent as a memento in return.

The Move Underground

Slowly during the 1950s, public opinion began to turn against atmospheric atomic testing. The contamination of a Japanese fishing vessel in 1954 became an international incident when the boat accidentally received fallout from the testing of a U.S. thermonuclear bomb in the Pacific. Reports of radioactive contamination of milk and fish from fallout began to panic American consumers and mothers with young children. Nobel Prize–winning chemist Dr. Linus Pauling, at the California Institute of Technology, cited the dangers of fallout for milk and led a group of 2000 prominent scientists in calling for an end to atomic testing. For his efforts, the government tried to cut off his grant funds for research. During the 1956 presidential election, Adlai Stevenson repeatedly raised the issue of fallout and atomic testing with American voters.

To fight the increasing calls for a ban on atomic tests, the staff of California's Donner Laboratory searched for a means to criticize Pauling's assertions. They liked the argument of Dr. John Gofman — who was then a proponent of nuclear power — that high-dose radiation results could not be extrapolated to low-dose radiation. Ernest Lawrence of the Lawrence Livermore Laboratory would later accompany Gofman to a University of California board of regents meeting to defend UC's sponsorship of the weapons testing program and downplay the dangers of low-level radiation.

In the late 1950s, the Soviet Union and the United States temporarily halted atmospheric testing, but within a few years the tests were resumed. By 1959, the U.S. government had established the Pacific Missile Range, starting at Vandenberg Air Force Base and Point Mugu Naval Base in southern California and ending at the Kwajalein atoll in the western Pacific, not far from Bikini and Eniwetok. In May 1962, a submarine submerged off the California coast launched a Polaris missile with a nuclear warhead that traveled 1250 miles down range to Christmas Island and exploded with a force of about 40 times that of Hiroshima. Five days later, an ASROC anti-submarine rocket with a nuclear warhead was fired from the deck of a ship and exploded underwater about 360 miles southwest of San Diego in the Pacific. In all, the U.S. exploded 89 nuclear devices during 1962, a one-year record.

The public and political will for a treaty to ban atmospheric weapons tests, and the technological means to monitor adherence to it, finally led the United States, the Soviet Union and Great Britain to sign a limited nuclear test ban treaty in 1963 that ended atmospheric testing by those countries. The U.S. tests then moved underground, where they have remained ever since.

The U.S. began experimenting with underground testing to reduce the radiation dangers from fallout as early as 1956. Regular underground testing in Nevada began in 1961 at the Nevada Test Site, a 1350-square-mile reservation that extends to within 20 miles of the California state line and Death Valley National Park. Since then, there have been more than 500 announced underground tests, some larger than one megaton in size. At least 20 of these have released quantities of radioactivity sufficient to be detected at off-site locations. At least five were detected in California, at monitoring stations at Death Valley Junction, Shoshone and Bakersfield.

The most serious off-site exposure came on December 18, 1970, from the accidental release of radioactivity for about 24 hours after the Baneberry test, an explosion 910 feet below the Nevada desert designed by the Lawrence Livermore Laboratory. While monitoring the radiation released from the test afterwards, officials found 413 contaminated automobiles close to the test site and measurable increases in the radioactive levels in milk at five locations outside Nevada, including Bakersfield. At least 86 people at the site were exposed by the Baneberry test and had to be decontaminated. Three later died from leukemia, and legal actions have been filed seeking compensation.

After Baneberry, the federal government undertook a major effort to reduce the risk of accidental releases of radiation from the test site. In 1978 a Department of Energy official testified before Congress that "it is unlikely that another major venting will occur." But on September 25, 1980, radioactive gas from the Riola test began seeping into the air and was detected within 10 miles of the California border. Department of Energy spokesman David Jackson described this as the "first accidental release of radioactivity at the test site since November 1971," although there had been a number of "very small controlled releases" since then.

In 1979, an interagency federal task force suggested exploring an increase in "air, soil and water monitoring to detect fallout from tests conducted by other nations or from an accidental venting from U.S. underground test locations. . . ." The group also recommended the development of response efforts that included the "decontamination of food, evacuation or sheltering of persons." On April 30, 1981, Lawrence Livermore Laboratory conducted the fourth announced underground test for the year and the 469th announced test in Nevada.

Today's political climate has changed remarkably from the unquestioning attitude of public officials during the chilliest days of the Cold War. Governor Jerry Brown issued a statement in 1979 that he was "shocked" the Navy had "covered up" the incidents of radiation contamination at operations Wigwam and Swordfish right off the coast of San Diego. Brown more recently has led a fight among the regents of the University of California to reduce its management of Livermore and Los Alamos nuclear weapons labs for moral reasons. Scripps' deputy director Jeffrey Frautschy concedes that changes in public feeling towards radiation and nuclear testing since the 1950s would prevent another test like Wigwam. But for some it is too late. In December 1979 the USS Eastman's Captain Ross Pennington died of leukemia, 23 years after his ship came home to San Francisco — and one year after the Veterans Administration awarded him disability for his repeated exposures to radiation at Wigwam and Redwing.

THE DAY THE MISSILES HIT SAN DIEGO

PAUL NUSSBAUM

Somewhere in the Soviet Union, there is a nuclear missile with San Diego's name on it.

San Diego now is one of 61 places in the United States, seven in California, considered by defense planners as most likely to be attacked first in a nuclear war.

Because of its nuclear submarine repair shops, nuclear weapons storage sites and its concentration of Navy ships and installations, San Diego was designated on December 31, 1980, as a "counterforce" site — an area vital to the United States' ability to wage nuclear war.

With detente in decline, the Strategic Arms Limitation Talks in limbo, Soviet soldiers in Afghanistan, fears of another "Vietnam" in El Salvador and tough talk at the White House and the Kremlin, the world today ponders more closely the prospect of nuclear conflict.

Any discussion of the effects of nuclear war involves enormous uncertainties. No one knows exactly where the enemy missiles would strike or how much destruction they would wreak. Much would depend on the time of day and the weather, as well as evacuation efforts and the condition of shelters and medical facilities.

But interviews with half a dozen nuclear war and civil defense experts and a review of more than a score of documents, reports and books makes it possible to draw a likely portrait of San Diego in the event of nuclear attack.

This fictionalized scenario deliberately eschews the sanitized Pentagon terminology that obfuscates reality — terms like "mutually assured deterrence," which means the ability of the United States and the Soviet Union to blast each other off the face of the globe, and "widespread nuclear exchange," which translates into millions of charred bodies and broken buildings.

After its journey of 6000 miles, the Soviet SS-18 missile is terrifyingly accurate. Flashing in from the north, it drops to 6000 feet and explodes directly over the Coronado Bridge. Although much of San Diego and the surrounding area has been evacuated in the days and hours before the explosion, about 350,000 people are still within eight miles of ground zero. Of those, 135,000 are killed almost instantly and a like number seriously hurt.

About 63,000 people in that eight-mile radius (stretching roughly from Imperial Beach to Mission Bay and inland to San Diego State) and another 10,000 people three miles farther out will die painful deaths from their burns within days. Thousands more will suffer lethal doses of radiation poisoning from the blast and will die within weeks. Many others will have their physical resistance to disease so drastically lowered by radiation exposure that they will fall victim to illnesses they normally would have survived. And survivors will face a lingering danger from radioactive fallout.

A Blinding Flash

The one-megaton bomb (a small bomb in the modern nuclear arsenal, but still 67 times more powerful than the Hiroshima A-bomb of 1945) explodes on a clear Thursday morning, at about 11 o'clock. The blinding flash is the last thing drivers crossing San Diego Bay ever will see. Instantly, the gracefully arched bridge vanishes in a fireball that, at its core, is hotter than the sun. Within a radius of about two miles, the heart of San Diego disappears as the first flash of heat is followed immediately by a shock wave exerting more than a million pounds of pressure and then, more slowly, by a spreading cloud of radiation. Most of the ships floating at the 32nd Street Naval Station are ripped apart and thrown in chunks into the rubble that was National City and Logan Heights. The Hotel del Coronado is flung into the Pacific as splinters and dust. So are the hotel's guests.

Downtown, the modern steel-and-glass skyscrapers are shredded, their walls and occupants blown out by the pressure of the blast and the incredible shock wave that follows. City Hall doesn't so much crumble as explode, sending huge pieces of steel and concrete and smaller slivers of glass flying outwards at tremendous speeds. The people inside are tossed like dolls through the air by winds that are twice as ferocious as the most powerful hurricane. Their incinerated bodies are scattered throughout rubble that is distributed evenly in a layer many feet deep throughout the downtown area.

"America's Finest City" is suddenly a wasteland. The Bank of America, the stately County Administration Center, the luxurious Westgate Hotel are all gone, swept away in an instant.

From the Naval Amphibious Base on Coronado Island to the southern edge of Balboa Park, the population is burned to death by a wall of heat moving at the speed of light from the explosion. Those not incinerated are killed by the shock wave of air that hurls people and buildings and trees and cars into each other.

A woman in Hillcrest four miles from ground zero is running from her house to her car at the moment of the blast. Five seconds later, she is burned to death. Pressure waves follow in another four seconds, subjecting her charred body to the same forces a diver would encounter about 20 feet below the surface. Then winds of 180 miles an hour throw her lifeless body, along with the rubble that was her house, into an overpass abutment. Shards of glass and steel and wood turn the corpse into a grisly pincushion.

Within the 2.5-mile radius around ground zero, virtually no one survives the blast. The initial fireball is followed by a shock wave that creates winds of more than 2000 miles an hour at ground zero (by comparison, winds of 74 mph constitute a hurricane, and the most powerful hurricanes produce winds of less than 200 mph). The winds drop off quickly — three miles from ground zero they are about 290 mph, six miles away they are 95 mph, and 11½ miles out, 35 mph.

Planes taking off from Lindbergh Field are batted to earth like insects, tossed into the rubble that was the Marine Corps Recruit Depot.

Even reinforced concrete buildings are leveled. Four miles out, including the Hillcrest area where the woman met her fate, about half of the people are killed outright, incinerated, blown into buildings and crushed, and another 40 percent are injured as houses and commercial buildings collapse on them.

Eight miles from ground zero, reaching into Lemon Grove, Chula Vista, Mission Bay and Tierrasanta, at least five percent of the people are killed and 45 percent injured by the blast; homes are destroyed or irreparably damaged, commercial buildings are severely damaged, and the winds are still strong enough to blow people out of modern office buildings. This is where there is the highest danger of uncontrollable fires, as the heat from the blast ignites drapes, upholstery and paper inside the houses, and still-standing buildings provide fuel for spreading the blazes. People who are outside when the bomb explodes suffer blistering second-degree burns, and their clothing catches fire. Many of them will later die.

San Diego suddenly has tens of thousands of burn victims requiring specialized hospital treatment, but most won't get it. There are only 85 specialized burn centers with about 2000 beds in the entire United States. And in San Diego, two-thirds of the hospitals have been destroyed or severely damaged.

Blunting a Counterattack

The bomb that has devastated San Diego is among the first to hit the United States. Minutes earlier, a wave of missiles struck intercontinental ballistic missile silos in the Dakotas, Arizona, Kansas and Arkansas, followed by a rain of nuclear destruction on Strategic Air Command bases throughout the nation. These attacks on U.S. "counterforce" sites were designed to prevent the Americans from responding with an obliterating counterattack.

(Defense analysts say most missiles and bombs would reach their targets, and the most accurate would strike within 600 feet of the target. The United States has no anti-ballistic missile systems deployed — the ABM treaty of 1972 limited the United States and the Soviet Union to two ABM defense sites each; one to defend a missile site and the other to shield the nation's capital. But none have been installed.)

The San Diego attack is more of the same — designed to wipe out the United States' ability to mount a return attack. If the United States is to be stopped from fighting back effectively, San Diego must be destroyed.

If the Soviet attack were restricted to U.S. counterforce targets — which some analysts believe is the least irrational way of waging strategic war — a single bomb near the heart of San Diego's Navy complex could be expected. If, however, there is what planners euphemistically refer to as "widespread nuclear exchange" — all-out nuclear war with each side launching massive attacks on military, industrial and civilian targets, all of San Diego County is a high-risk area. In such a war, 80 percent of the county's population that was not moved east to safer areas probably would be killed, civil defense planners estimate. Furthermore, there would not be just one bomb for the county, aimed primarily at Navy targets near downtown. A whole cluster of nuclear bombs likely would descend on Southern California.

In San Diego County, everybody between Solana Beach and Dulzura, Rancho Bernardo and Alpine, Fernbrook and San Ysidro would be in harm's way. Another bomb would be earmarked for Camp Pendleton, the home not only of Marines but also, according to the Center for Defense Information in Washington, of nuclear weapons. The Marines decline to confirm or deny the reports of weapons storage.

Even those not blown apart, burned to death or crushed under crumbling structures would be threatened by radioactive fallout. The entire county would become a radioactive hot spot. A local scientist who helped prepare the federal Office of Technology's extensive "Effects of Nuclear War" says that if San Diego County is attacked in an all-out war, "It will be like those posters said 15 years ago — bend over and kiss your ass goodby."

The Coronado Bridge blast and the attacks on other U.S. counterforce sites had come after weeks of mounting tensions between the United States and the Soviet Union. But the leap from unpleasantness to war had been dizzyingly swift. U.S. authorities had anticipated they would have almost a week to make an orderly evacuation of the nation's cities. They were wrong.

The most fearful and wisest among San Diego's population began to "spontaneously evacuate" the weekend before the attack. They put the kids and some groceries and sleeping bags in the family car and drove east. On Wednesday, the day before the attack, the Soviets had begun a mass evacuation of their own cities, triggering an order by the U.S. president to evacuate American cities as well.

Southern California had worried evacuation planners for years — the area would be the most difficult in the nation to relocate, because there are so many people and so few safe places to put them. Their fears were borne out in San Diego County when the panic evacuation started. Interstate 8 became clogged with people trying to flee, and those without cars tried desperately to get rides, sometimes at gunpoint. Wrecks were common, adding more human flotsam to the waves of people who literally were running for the hills.

By Thursday morning, orbiting satellites report activity at Soviet missile bases that could mean only one thing: nuclear warheads — and nuclear war — are about to be launched. Minutes before the first missiles are propelled from their underground silos, the White House issues an emergency warning — it's too late to run; take cover. But in San Diego, there are pitifully few places to hide. Those who have basements crawl into them. Others make the fatal mistake of taking refuge in downtown buildings marked as fallout shelters, buildings that won't be there when the fallout arrives.

A last-minute flotilla of sailboats and pleasure craft tries to escape by water, but it is turned back by Navy patrol boats trying to keep the lone exit from the harbor open for the parade of warships heading out to sea. Weekend sailors with boats normally moored in Mission Bay are more fortunate; many scurry out the entrance channel to a dubious safety on the open ocean.

In the final few minutes before 11 o'clock, the terror and confusion is heightened when electric power fails suddenly. San Diegans don't know it, but the Soviets have just exploded several nuclear bombs 20 miles above the West Coast, sending out an immensely powerful wave called electromagnetic pulse (EMP). This massive pulse of energy, similar to the electricity from lightning but with a rise in voltage 100 times faster, knocks out American over-the-horizon radar and ravages San Diego Gas & Electric's power grids.

Lights go out. Elevators stop in their shafts, trapping hundreds of people trying to flee office buildings and apartment complexes. Radio and television signals are drowned in a sea of static. Sensitive civilian and military computers have their electronic brains fried.

At the moment of the blast, the bomb that devastates San Diego also creates deadly radioactive fallout (although much less than if the bomb had been exploded on the ground) that begins to settle on the devastated downtown areas within 10 minutes. There it has little effect because almost no one survived the initial explosion.

As the mushroom cloud spreads eastward, though, it begins to rain radioactive particles on the terrified survivors. People left without shelter by the blast within about 12 miles of ground zero could be killed within a week by the levels of radiation they absorb. Many of the most harmful particles decay rather quickly, but winds out of the west carry dangerous levels of fallout into Imperial County.

A Shattered Society

Some people beyond the 12-mile radius sicken and die, but most recover from the short-term effects. The long-range implications for the survivors are more grim: They are much more likely to get cancer within a few years, more likely to have miscarriages, more likely to have deformed children.

The ability of San Diego to fend for itself and to care for its injured has been destroyed by the attack. Two-thirds of the area's hospitals have been destroyed or heavily damaged by the blast, and the remainder are in the path of radioactive fallout.

Fire crews that survive the blast with any workable equipment concentrate their efforts in the close-built beach communities where the fires are jumping from house to house and in the new housing tracts north of Interstate 8. The dry undergrowth that makes San Diego County a tinderbox for summer brush fires feeds the spreading holocaust.

Rescue efforts in the hours after the attack are directed from a bunker in Fletcher Hills about 11 miles from ground zero. About 60 Civil Defense officials had taken cover there before the blast, and they are well-protected from the fallout. They have food and water for two weeks, fuel and radios. But they don't have the county's leaders — no plans were made to save the mayor or the city council members or the county supervisors — so much of the knowledge that would be vital in restoring services and order has perished.

As the survivors try to recover, tales of heroism and villainy abound. Neighbors rescue neighbors from collapsed buildings and burning rubble; outmanned doctors, firefighters and police officers work without rest to find and treat the injured. Teen-age gangs commandeer cars stopped in traffic, beating and robbing the drivers. Homeowners lucky enough to still have homes shoot at the less fortunate who come looking for help and refuge.

Eventually, nearly everyone who survived the blast is moved into overcrowded "host" areas set up in Imperial County. The economy of the nation, not to mention San Diego County, is shattered by the attacks, and the first task the grieving survivors face is staying alive. Finding food and water that is uncontaminated by radiation is a major chore for harried officials working with unprecedented "emergency powers" that amount to martial law.

Whether San Diego ever will be rebuilt is an open question. Areas of the city nearest ground zero won't be safe to inhabit (at least by peacetime radiation standards) for several years. What is clear, though, is that whatever rises from the ashes will be much different from the city that existed on the day that nuclear war came to San Diego.

U.S. Navy

PART FOUR

A NUCLEAR ATLAS OF CALIFORNIA

California plays host to nuclear materials of every kind: weapons stockpiles; research, commercial and naval reactors; large industrial, medical and scientific laboratories; and an assortment of dump sites offshore. These materials are concentrated in and around the state's largest metropolitan areas.

What follows is a series of eight maps put together by the staff of the Center for Investigative Reporting. This small atlas details the extent to which radioactive materials have become part of our environment. We've added the routes by which high-level waste is shipped, the defense contractors which manufacture nuclear weapons and the complex web of active earthquake faults which underlies the state's nuclear facilities. They are part of the far-reaching impact on a state in which more than 2000 facilities are licensed to handle radioactive materials.

Photo: Navy helicopters like this one in San Diego transport nuclear weapons

MAJOR RADIOACTIVE MATERIAL SITES •
AND EARTHQUAKE FAULTS   ------·

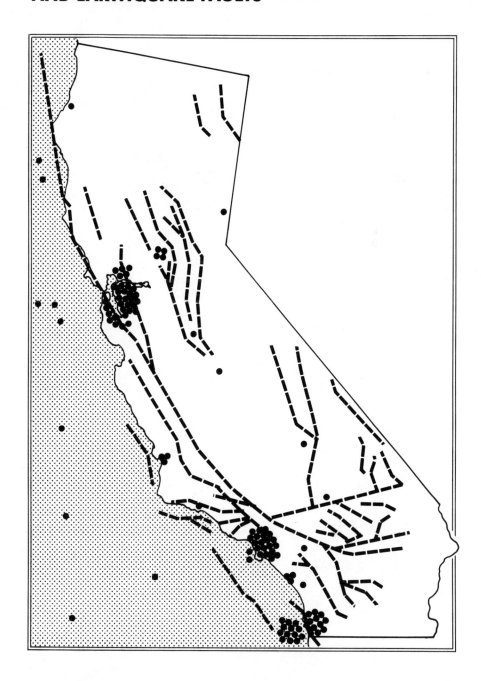

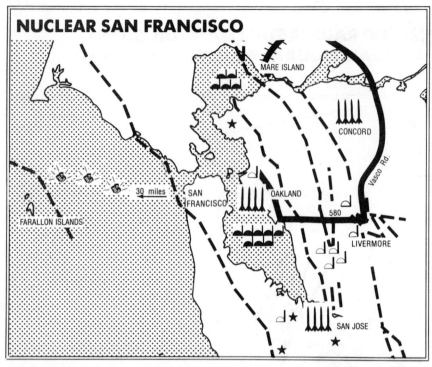

NUCLEAR SAN FRANCISCO

MARE ISLAND

CONCORD

Vasco Rd.

FARALLON ISLANDS

30 miles

SAN FRANCISCO

OAKLAND

580

LIVERMORE

SAN JOSE

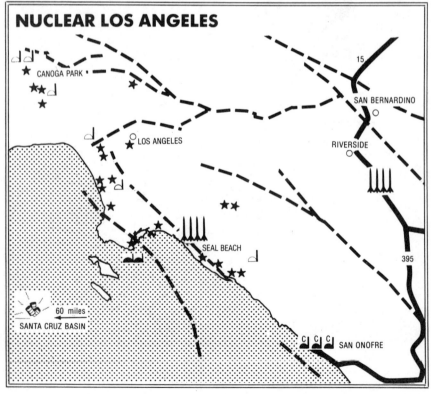

NUCLEAR LOS ANGELES

CANOGA PARK

15

SAN BERNARDINO

LOS ANGELES

RIVERSIDE

SEAL BEACH

395

60 miles

SANTA CRUZ BASIN

C C C SAN ONOFRE

◀ If you live in the Bay Area, chances are you live in a nuclear neighborhood. Barges carrying atomic weapons and radioactive waste plow through the busy harbors. Helicopters lift A-bombs over peaceful suburbs. As many as 22 military and civilian reactors sit atop active earthquake faults. The water supply for more than a million East Bay residents runs within a mile of the Concord nuclear ammo dump. Is this any way to run a city?

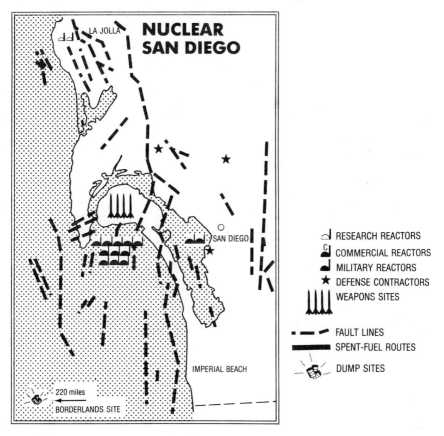

▲ The nuclear navy is omnipresent in San Diego, home port to huge nuclear-powered cruisers, numerous submarines and a nuclear arsenal located a mile from downtown. The navy repeatedly has spilled radioactive coolant into the bay, and regularly transports nuclear weapons on city streets. Official reactions are restrained because of the community's heavy dependence upon military spending.

◀ Metropolitan L.A. is home to as many as 11 military and civilian reactors, two major atomic weapons arsenals, and one of the largest concentrations of nuclear arms contractors in the nation. The San Onofre Nuclear Generating Station, about 50 miles from downtown L.A., sits in one of the most active seismic regions in North America. A core meltdown there could contaminate 16,000 square miles, kill nearly one-half million people and cause $180 billion in damages.

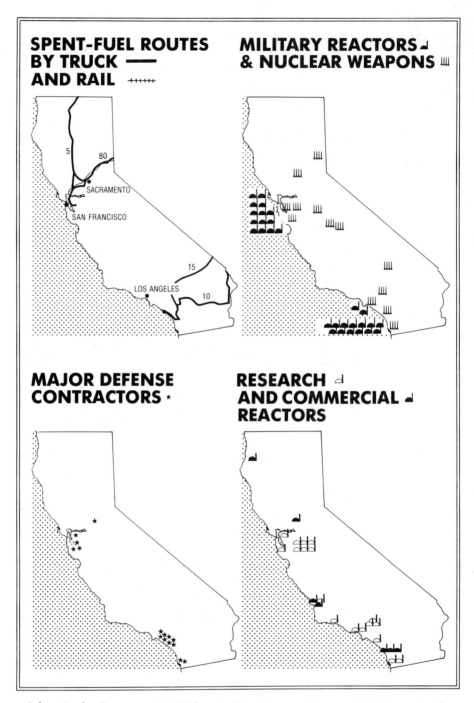

SPENT-FUEL ROUTES BY TRUCK —— AND RAIL ++++++

5
80
SACRAMENTO
SAN FRANCISCO
15
LOS ANGELES
10

MILITARY REACTORS ⌐ & NUCLEAR WEAPONS ⊞

MAJOR DEFENSE CONTRACTORS *

RESEARCH ⌐ AND COMMERCIAL ⌐ REACTORS

Information for all maps was compiled from the following sources: Reactors—U.S. Departments of Defense and Energy, Nuclear Regulatory Commission; Weapons—Center for Defense Information, Center for Investigative Reporting; Other Radioactive Sites—U.S. NRC, California Department of Health Services; Spent Fuel Routes—U.S. NRC, Center for Investigative Reporting; Defense Contractors—*Aviation Week and Space Technology*, U.S. DOD; Fault Lines—U.S. Geological Survey, California Division of Mines and Geology, *Los Angeles Times*, Earth Science Associates/City of Concord.

PART FIVE

SCIENCE AND INDUSTRY IN THE NUCLEAR STATE

Billions of dollars are spent annually in California to tap the power of the atom for science, industry and the national defense. Taken together, these efforts may constitute the largest single industry in the state, surpassing even agriculture and tourism. As the next four articles show, atomic bombs, fission reactors, accelerators and other nuclear equipment are all designed and constructed within the state. They have given California an estimated 55 military and civilian reactors, an assortment of plutonium laboratories and the nation's largest defense industry.

As taxpayers and as rate payers, we have financed these industries. But as with so much of Nuclear California, we have also just begun to tally their social, economic and environmental costs.

Photo: A worker completes assembly of a guided missile at Hughes Aircraft Co.

ATOMIC ASSEMBLY LINES

MARK SCHAPIRO AND MARC BEYELER

Poised on the California coast, an hour's drive northwest from Santa Barbara, sits America's nuclear launch pad. From here, at Vandenberg Air Force Base, the deadliest of America's nuclear missiles are test-fired 4000 to 6000 miles into the western Pacific. Propulsion systems are tested for their ability to deliver nuclear warheads on target.

The latest flurry of activity at Vandenberg revolves around the MX missile system. Facilities currently under construction will provide the experimental testing grounds for a series of launches scheduled for early in 1983. Each of the 10 warheads per MX missile is 26 times more powerful than the bomb dropped on Hiroshima, and three times more powerful than the previous generation of Minuteman missiles.

Construction crews with bulldozers are readying the launch site and building what is known as the "designated assembly area," where the various components of the missile are to be assembled after being shipped from factories throughout the country. The bulk of the shipments will not have to travel far, however. Most of the project's largest contractors are operating in California: Rockwell International, Aerojet General, Martin Marietta, Northrop and TRW, Inc. The guidance and control system will come from Anaheim; the second rocket stage from Sacramento; and the fourth rocket stage from Canoga Park. All of the companies involved have learned to cooperate in a state whose economic vitality depends in large part on America's military presence around the world.

In addition, a national map of nuclear weapons research facilities compiled by the American Friends Service Committee shows California has the heaviest concentration in the country, with 26 of the nation's 61 national corporate contractors. With a bizarre litany of project titles — such as "Husky Pup," "Hussar Sword" and "Dice Throw II" — the contractors range from General Electric in Santa Barbara, studying how many people would be killed from a single nuclear blast, to Lockheed's mysterious "Novel Kill Mechanisms Program" in Palo Alto.

California averages a 20 percent share of national defense work, by far the largest of any state. Sixteen billion dollars in defense contracts came into the state in 1980. This figure does not include the $6.8 billion spent on military base operations here. The State Department of Resources reports that in 1980 defense contracting was almost as big as the state's two largest industries, agriculture and tourism. A series of congressional studies sponsored by Northeast and Midwest congressmen points out that California is one of only a few states that receive more federal money, in the form of defense contracts, than they pay out in federal taxes.

A little-noticed set of hearings by the California State Senate Select Committee on Investment Priorities and Objectives, held in 1978, painted a graphic picture of the role of defense spending in California's economy. The hearings revealed that more than five percent of California's work force is directly dependent on the Defense Department for its livelihood. Excluded from that figure are NASA and Department of Energy employees involved in nuclear weapons–related work. One-third of California's military work force is employed by private contractors, and two-thirds are direct civilian or military employees of the Department of Defense. Add to this "the multiplier effect," a mathematical progression which according to most economists dictates that two additional jobs are created for each defense position, and the effect of defense spending in the state is even more pronounced.

Defense spending is highly concentrated within the state, with more than three-quarters of California's Pentagon expenditures centered in Los Angeles, Santa Clara, San Diego and Orange counties. Although Los Angeles captures the greatest number of defense contracts — $6.1 billion worth in 1980 — Santa Clara has the highest per capita defense spending in the state and one of the highest in the country — over $1000 in military money for every person in the county. A Stanford Research Institute study released in 1980 concluded that eight percent of the county's residents are directly dependent on defense spending.

President Reagan's proposed $1.8 trillion increase in the defense budget during the next five years will substantially add to the state's share of military contracting through the 1980s. David McFadden, who has been tracking California's defense industries for six years with the Mid-Peninsula Conversion Project in Mountain View, explains, "One-quarter of Reagan's increase in procurement funds — money for buying weapons — will go to California, mostly for accelerated work in nuclear and electronic warfare technology and increased purchases of aircraft and ships, which are California's specialties."

The primary recipients of these new contracts will be the 10 top contractors who have traditionally accounted for approximately 50 percent of defense contracts in the state (see Table I).

California's own military-industrial complex is a labyrinthine web of prime and sub-contractors that compete and sublet among themselves. The Defense Department's guide to California contractors — "Prime Contract Awards Over $10,000 by State, County, Contractor and Place (FY 1980)" — runs 54 pages, 50 companies per page, listing a total of 7700 companies with a direct stake in military weapons production.

TABLE I: TOP 10 CALIFORNIA DEFENSE CONTRACTORS IN 1980
          (Figures in thousands)

		NATIONAL RANK
1. Hughes Aircraft	$1,568,470	7
2. Lockheed	1,516,051	6
3. General Dynamics	833,510	1
4. Rockwell International	609,977	14
5. FMC	493,167	17
6. TRW	414,739	27
7. McDonnell Douglas	395,063	2
8. Ford Aerospace	375,317	34
9. Northrop	354,111	11
10. Chevron USA	326,925	30

Sources:

Aviation Week and Space Technology 4/27/81. National rank only.

Department of Information, Operations, and Reports, U.S. Department of Defense — FY 1980; "Prime Contractors Over $10,000 — California."

The close relationship between major contractors and the military is, in some cases, illustrated by their physical proximity. At Norton Air Force Base near Riverside, for example, TRW executives in charge of planning for the MX literally share office space on base with Air Force Ballistic Missile Office personnel. (TRW has been, since the 1950s, the systems engineer for all major American nuclear missile programs, including the Atlas, Titan, Minuteman and now the MX.)

Since the mid-1970s the development of California's military economy has been heavily oriented toward production of a "new generation" of nuclear weapons. This fresh stream of weaponry arises out of a change in Pentagon strategy that now envisions a "winnable" nuclear war — a shift away from the concept of Mutual Assured Destruction (MAD) in which both sides were theoretically deterred from launching an attack from fear that neither could emerge as a "winner." Now new weapons are demanded by the Pentagon that are theoretically both accurate enough to pinpoint enemy military targets and powerful enough to knock them out. Ironically these weapons, though not designed to be aimed at cities, exacerbate the existing arms race and threaten future arms control agreements by introducing the possibility of attacking military targets only, making a "limited" nuclear war an increasingly attractive strategic option.

The major components of these "new generation" weapons are the MX missile, the B-1 bomber, the submarine-launched Trident missile and the Cruise missile, which can fly under its own power to within a hundred feet of a target thousands of miles away. California arms producers hold a considerable stake in this revised version of what its critics call "limited Armageddon" (see Table II).

TABLE II: MAJOR CALIFORNIA CONTRACTORS FOR
"NEW GENERATION MISSILES"

MX	TRIDENT	CRUISE
TRW *Redondo Beach &* *Vandenberg Air Force* *Base*	Lockheed Missiles & Space Company *Sunnyvale*	Garrett Airesearch *Long Beach*
	Data Design Lab *Cucamonga*	General Dynamics *San Diego*
Martin Marietta *Vandenberg Air Force* *Base*	Datron Systems *Los Angeles*	Litton Industries *Woodland Hills*
Rockwell International *Anaheim, Canoga Park*	General Dynamics *San Diego*	Lockheed Missiles & Space Company *Sunnyvale*
Aerojet General *Sacramento*	Hughes Aircraft *Culver City*	Marquardt Company *Van Nuys*
Northrop *Hawthorne*	Interstate Electronics *Anaheim*	Rockwell International *Canoga Park*
	Rockwell International *Anaheim*	United Technologies *Sunnyvale*
	Singer Company *Glendale*	GTE *Mountain View*
	United Technologies *Sunnyvale*	
	Westinghouse *Sunnyvale*	

Sources:
National Action/Research on the Military Industrial Complex (NARMIC), American Friends Service
Committee
Council on Economic Priorities
Corporate Information Center, Interfaith Center on Corporate Responsibility

What may be the most potentially devastating addition to this arsenal
are space weapons, which received a $50 million appropriation for research
and development in 1982. Lawrence Livermore Laboratory, early in 1981,
successfully conducted tests on a nuclear-powered laser weapon that would be
based on orbiting space stations.

California's traditional lead in aerospace technology will ensure the state
a primary stake in the development of this new level of weaponry.

Researching Nuclear War

Critical to California's military-industrial complex are the research facilities that make it their business to challenge and test concepts of war and
weaponry, fueling the new ideas that propel the arms race.

The Department of Energy is responsible for nuclear weapons research
and development in this country. President Reagan's 1982 budget marks a

turning point in the agency's history, the first year in which more money will be allocated to defense-related activities than to energy development.

The Department requested $5 billion — or 36 percent of its budget — for nuclear weapons work alone, noting at a congressional budget presentation that "most of the increase is due to new requirements for nuclear weapons and nuclear materials, as outlined in the President's recent 'Nuclear Weapons Stockpile' memorandum." The Department's budget request does not even include nuclear-weapons research, such as development of the MX power system, which had been proposed for funding through the "solar and conservation" budget.

At the pinnacle of scientific inquiry into nuclear weaponry is, of course, the University of California, which operates Lawrence Livermore National Laboratory, 40 miles from San Francisco, and Los Alamos Scientific Laboratory in New Mexico. These two labs are responsible for conceiving and designing every nuclear warhead in the American arsenal.

Military Spending for Whom?

Ironically, President Reagan's program for increases in defense spending — larger boosts than occurred during the Vietnam War buildup in the 1960s — threatens his own economic recovery plan. Already, federal deficits in 1983 and beyond are estimated by executive and congressional advisers at more than $100 billion, yet the President continues to seek massive military spending increases. Numerous economists also have warned that the "boom" that inevitably accompanies military spending — and from which California is expected to benefit — has serious adverse consequences for the economy. Economists such as Seymour Melman at Columbia University and Lester Thurow at MIT point to increased inflation, lower productivity and fewer jobs than equal investments in civilian industry would provide as the side effects of military spending.

No less an expert on military contracting than Simon Ramo, a co-founder of TRW and a science and technology adviser in the Reagan Administration, wrote in *America's Technological Slip* (published in 1981) that, "on the whole [American technological development] has been hurt rather than helped by our heavier involvement in military technology as compared with other [Western] nations."

Congressman Les Aspin (D-Wis.) released a study in April 1981 which documents the decline in industrial productivity throughout the late 1960s and into the '70s resulting from federal defense spending. "When arms are produced at the leading edge of technology," Aspin commented, "they consume vast quantities of technical talent." He noted that defense and space industries employ up to 50 percent of the scientists and engineers engaged in research and development work in the United States. Recently released figures from the Office of Management and the Budget indicate that this trend is accelerating under Reagan, as the new budget will raise the military's share of all federal research and development expenditures to 53 percent during the next two years.

Seymour Melman, a professor in industrial engineering at Columbia University and author of *The Permanent War Economy*, which describes the economic impacts of defense spending, comments: "For every $100 of new capital formation, $46 is spent on the military. This diverts resources that otherwise would be put into capital and investment costs for civilian production."

While consuming an enormous amount of the country's resources, military production does not produce jobs commensurate with the amount of investment. Employment Research Associates, an economic research firm in Lansing, Michigan, reported in a 1981 study that defense spending contributes to unemployment by diverting resources from private investment and consumption. Using Bureau of Labor Statistics figures, the study concluded that for every $1 billion spent on defense, the nation loses 10,000 jobs. According to the study, $1 billion spent in the civilian industrial sector would create 27,000 jobs; but the same $1 billion spent for the defense industry would create only 17,000 jobs.

The Council on Economic Priorities, a nationally recognized group of public interest economists in New York, undertook an even more precise study of the number of jobs displaced by a single defense project. The council zeroed in on a $1 billion investment in the MX missile system and compared it to similar investments in other areas of the economy. Energy conservation created 22 percent more jobs than the missile system; solid waste treatment 24 percent more, mass transit 49 percent more and child care over 126 percent more jobs.

At the same time, military spending leads to spiralling inflation because the money it generates does not produce goods that can be consumed. "In a classic economic sense," says Seymour Melman, "military goods are nonproductive, in that they can't be consumed. A tank cannot be sold in the marketplace."

President Reagan's Council of Economic Advisors maintains that economic recovery will be sparked through tightening the money supply and cutting domestic spending (making room for military programs). But, asserts Melman, military spending will exacerbate the current inflation by producing income without producing an equivalent amount of consumer products, putting more money into circulation than there are goods to consume.

While numerous economists have focused on the negative impacts to the national economy resulting from the proposed increases in defense spending, no similar studies have been done detailing the impacts on California. Though the state's defense industry may reap huge profits during the 1980s, Californians may feel the repercussions in lost civilian employment and investment and continuing high rates of inflation.

It is, indeed, the great irony of California's military-industrial complex that while helping to produce fully one-fifth of the nation's armaments, the state may find that its heavy reliance on military spending could undermine its economic stability.

"California may benefit from a Niagara Falls of cash," says Melman, "but it will show more and more output of an economically meaningless product, producing artifacts that may help to end the human race."

THE DEMISE OF THE NUCLEAR INDUSTRY

JIM HARDING

Imagine a coastline studded with 70 new nuclear power plants, including seven at Diablo Canyon, two reprocessing facilities and a breeder reactor near the Imperial Valley and uranium mining in the San Joaquin Valley near Bakersfield.

This vision is not something out of Disney's Tomorrowland. As recently as 1974, it was considered the most likely nuclear future for California by a range of experts including the state's Public Utilities Commission, Stanford Research Institute and the RAND Corporation. Only three years ago, gubernatorial candidate Evelle Younger ran on a platform which endorsed this level of nuclear development for California. Even today, essentially similar nuclear forecasts are still the rule at former Governor Pat Brown's powerful lobby, the Council for Environmental and Economic Balance (CEEB).

Despite the staunch support of advocates like CEEB, however, the nuclear industry is on the ropes. The industry argues that it needs 30 new power plant orders annually to survive, but the last time *that* happened was 1973. Since then, cancellations have outnumbered new orders. In California, the two largest utilities say they will not build any further nuclear plants beyond those under construction—not because they don't like the plants, but because they don't need them. A combination of energy conservation programs, industrial co-generation efforts and renewable resource projects has been successful enough to render the overpriced and inefficient nuclear plants obsolete. Nationally, a similar trend is evident. The investment firm Merrill Lynch is "bullish" on nuclear cancellations, recommending that 18 nuclear projects be dumped to improve the financial health of their parent utilities.

This evidence of commercial disenchantment hasn't convinced President Reagan, gubernatorial candidate Mike Curb, the politically potent CEEB or the remnants of the U.S. nuclear industry that reside in California. The pro-nuclear stance of these groups apparently has become more strident with time. In 1978 Evelle Younger, then state attorney general, commissioned hydrogen-bomb designer Edward Teller to prepare Younger's California energy platform. In it, Younger and Teller urged California to "immediately embark" on a massive shift to nuclear energy by building 50 to 70 new nuclear plants, spaced an average of every 18 miles along the coast—and costing 15 times the current wealth of all the state's utilities.

In that same year, Pat Brown's CEEB, representing business and labor leaders throughout the state, commissioned a similar study. This report claimed that unless nuclear and coal power were expanded, the state would face blackouts every other day; lose 175,000 jobs annually and $4.7 billion in industrial earnings; and pay $12 billion for unnecessary increases in the price of electricity. Alternative sources were brusquely dismissed: hydropower was "almost completely developed" and solar, biomass and wind were "not expected to make any significant contribution by the year 2000." As for energy conservation, the report noted "an extensive conservation effort — which has yet to be mounted — might reduce consumption in 2000 by 15 to 20 percent. . . ."

The council published another report in 1981 which suggested that California must have 50 new nuclear or coal plants by 2000. And a recent report for the Department of Energy by the Oak Ridge Institute for Energy Analysis recommended clusters of new nuclear plants at San Onofre, Diablo Canyon and Rancho Seco, plus a reprocessing plant and a breeder reactor near Blythe, along the Arizona border.

All of these reports and recommendations are based on the idea that we are about to run out of energy — an idea that has been used historically to justify massive public expenditures on nuclear plants. According to 1973 forecasts of energy use, California needed 22 nuclear power plants in operation by 1980 in order to avoid daily blackouts. But 1980 came and went, and we had neither the plants nor the blackouts. California today has three nuclear plants and more blackouts caused by squirrels on transmission lines than by a lack of nuclear power. The state used 26 percent less energy in 1980 than projected in 1973, surpassing almost without effort the estimate CEEB said might be possible by 2000. And the potential for further improvement in conservation has barely been tapped.

California leads the nation in solar collector sales and has numerous small- and medium-scale hydropower units under construction. Huge windmill farms are planned, with extensive corporate backing; they will have sufficient capacity to dwarf present and projected nuclear capacity. Wood fuel, which was virtually untapped as an energy source in 1973, may now deliver more energy to U.S. consumers than nuclear power.

Even more evidence is supplied by a recent federal study which the Reagan Administration attempted to suppress. The five-volume report by the Solar Energy Research Institute shows how a "least cost" energy strategy to the year 2000 could increase GNP by two-thirds, reduce total energy by one-fourth, begin to phase out nuclear power and reduce non-renewable energy use by half — all with no lifestyle changes except those associated with stronger economic growth.

There are those who cling to the "Nuclear Can Do It All" philosophy, however, and they can be found throughout the New Right, the Reagan

Administration and in such conservative "think tanks" as the Heritage Foundation and CEEB. Their views are not founded on free market theory, because the marketplace has already done what nuclear critics alone could never do: pull the financial rug out from underneath the atomic industry.

According to utility analysts at Moody's, Standard and Poor's and Merrill Lynch, nuclear power no longer makes economic sense. Standard and Poor's derates the bonds of utilities with significant "nuclear exposure," and many utility analysts now argue that the risk of another Three Mile Island is too grave, that electricity demand is too soft, and that alternative sources are opening up too fast to risk huge investments in the future of nuclear power.

There is only one way the nuclear "option" can be resurrected and that is to emulate the state energy monopolies in France and the Soviet Union. In France, the state utility can seize any land it wants, build any plant it wants and charge any price it wants. It does not have to compare the cost or risk of its projects with conservation, co-generation, coal, wind or small hydro or geothermal plants. It is not required to seek the support of local governments or to hold a single public hearing on any aspect of its programs. Utilities in the Soviet Union are even less democratic. In both these nations, the nuclear decision has been made, and the central government is charged with carrying it out.

We could get that kind of "predictable yield" from nuclear power in California by eliminating all power plant siting processes; by guaranteeing any utility a rate of profit on all nuclear plants it builds, whether they are necessary, or operate safely, or operate at all; and by prohibiting states from rejecting nuclear plants for reasons of reliability, public health and safety, economics or competing land use. That future sounds unlikely, but there are disturbing signs that it could yet occur.

One well-placed proponent is Edwin Meese, President Reagan's top advisor and a former Alameda County district attorney. Before going to Washington with Reagan, Meese was directly involved in litigation, still ongoing, that would prevent any city or state government from turning down a nuclear plant proposal for any reason whatsoever, even if that plant were to be located in downtown San Francisco. Should that lawsuit eventually succeed on constitutional grounds, the battlefront will shift to Congress.

Until and unless the government intervenes, however, the economics of nuclear power will continue to drive the industry out of business. And one of the significant factors in the negative balance sheet for nuclear power is the impact of the anti-nuclear movement. Ten years ago, there was no anti-nuclear movement to speak of in California or in the nation. It was the industry's heyday; more plants were in operation, under construction and on order than there are today. Hardly a voice had been raised about the risks of nuclear energy. But in 1971 and 1972, scientists John Gofman and Arthur Tamplin warned of potentially high radiation risks from lax nuclear power standards, and MIT physicist Henry Kendall called attention to the risk of an uncontrollable nuclear accident. These critics initially had a tough time convincing the public they had anything worth listening to. The Sierra Club still officially endorsed atomic energy and even helped PG&E select the Diablo Canyon site.

The California Nuclear Safeguards Initiative, written by activist Alvin Duskin in 1973, brought the issue to the public, but the first efforts to carry the petition around the state met with dismal failure. In the five months that the secretary of state allows for petition circulation, a small group of dedicated volunteers managed to find 30,000 people who would sign their petition. They needed 16 times as many.

They went back to the public a second time, however, and with interest in energy and nuclear power climbing, got nearly half a million signatures and a spot on the June 1976 ballot. That initiative — Proposition 15 — lost, but political pressure to offer an alternative was so intense that, on the eve of the referendum, the state legislature passed a set of nuclear safeguard laws. One of those hit directly at the Achilles' heel of the nuclear industry by specifying that no more nuclear power plants could be built in California until a permanent method of radioactive waste disposal was demonstrated. Once the state made it clear that the law would be enforced, it amounted to a virtual ban on new plant construction. It is this set of safeguard laws that Ed Meese's political allies are now challenging in federal court.

The organizers of the Nuclear Safeguards Initiative created a major new direction for the environmental movement in California. During the late 1970s, tens of thousands of demonstrators gathered in front of the Diablo Canyon and San Onofre nuclear power plants. Governor Jerry Brown, recognizing that a powerful political force had emerged, became an outspoken critic of nuclear power. The lengthy legal challenges and protests against new reactors made it clear to California utilities that nuclear power could no longer be easily sold.

The fate of commercial nuclear power has been most damaged, however, by its own inefficiency as an energy source. Despite the best efforts of planners of CEEB, the Department of Energy and the Stanford Research Institute, utilities are finally realizing that the future of nuclear power is practically dead, and that an economy based on safe, renewable energy sources is entirely possible — and increasingly probable.

LAWRENCE LIVERMORE LABORATORY

MARCY DARNOVSKY

Hidden among the rolling hills of Livermore Valley, 20 miles east of Oakland, is a small modern fortress guarded by one of the largest police forces in Northern California. At first glance, it is an unimposing sight: several hundred acres of brick buildings with an almost academic innocence. But inside are the spawning grounds for half the nation's nuclear arsenal, in a place that could, according to its critics, unleash an uncontrollable nuclear accident.

The fortress is Lawrence Livermore National Laboratory (LLNL), a government complex staffed by 7200 scientists, technicians and laborers, with an annual operating budget of nearly $600 million. Stored within a single building at the lab are hundreds of pounds of weapons-grade uranium and plutonium. It is here at Livermore, and at the Los Alamos Scientific Laboratory in New Mexico, that government scientists have painstakingly designed every nuclear weapon in the United States.

In addition to its complex of office buildings and laboratories near the town of Livermore, LLNL runs a non-nuclear test range called Site 300 about 15 miles away, as well as part of the National Test Site near Las Vegas, where an estimated 20 nuclear bombs are exploded underground annually. A weapons laboratory operated by the Sandia Corporation is located just across the road from the Livermore complex.

The Dr. Strangeloves Next Door

The weapons labs take pride in their role as the nation's nuclear weapons brain trust. Lawrence Livermore's award-winning public relations office regularly churns out pamphlets featuring photos of shiny machinery, dedicated scientists and a golden sunset silhouetting a soldier next to the latest in improved nuclear weaponry.

There is a great deal, however, that officials at the lab don't readily reveal. All of the weapons programs are highly classified. The exact amount of plutonium and other radioactive materials it stores is also a secret — although it is licensed to handle up to 495 pounds of plutonium at any one time. Despite criticism from local congressmen and the Department of Energy (DOE), lab officials have played down or denied health, safety and environmental hazards posed by the work there. And though they boast of the lab's technical importance in researching and developing nuclear weapons, they don't publicize the critical role the lab plays in influencing U.S. defense policy.

The weapons labs are a powerful lobby. When the United States and the Soviet Union were negotiating a comprehensive ban on the testing of nuclear weapons in the late 1970s, the labs tried to stop it by conducting what the *Washington Post* called a "behind the scenes campaign" of "brass-knuckle bureaucratic warfare." The treaty was never concluded.

During the Carter Administration, Lawrence Livermore program director Roy Woodruff was quoted as saying, "There's been a gradual erosion [in the nuclear weapons program] and we're going to be very aggressive and out front at trying to turn that around." Los Alamos Director Harold Agnew went on to successfully lobby the administration to launch production of the neutron bomb.

Both Lawrence Livermore and Los Alamos are managed for the DOE by the University of California, an arrangement that has provoked vocal opposition from many UC students and faculty. A demand for severance has been raised by the UC Nuclear Weapons Labs Conversion Project, which argues that academia's good name is used to give a "cloak of legitimacy" to the labs and the scientists working there.

Similar arguments have been voiced by Governor Jerry Brown, and the controversy has reached the UC Board of Regents. But in June 1981 the regents voted to renew the DOE contract and keep the university in the nuclear weapons business for another five years. Brown's critics charge that the governor could have swung the vote the other way by filling three vacant seats on the Board of Regents with sympathetic appointees.

Research at the Lab

Lawrence Livermore was founded in 1952, largely at the urging of H-bomb designer Edward Teller, who felt that Los Alamos needed some competition to speed the dawning of the thermonuclear age. Since its founding, the lab has been responsible for the miniaturization of strategic warheads and the design of the Poseidon, Polaris and Minuteman ICBM warheads. It developed MIRVs (multiple independently-targeted re-entry vehicles) and MARVs (maneuverable re-entry vehicles) and pioneered the design of an anti-ballistic missile system that was rejected by Congress in 1972 as too expensive and too destabilizing (although it is now being reconsidered by the Reagan Administration).

More recently, the lab has been busy designing a whole new generation of bombs: warheads for the Lance and Cruise missiles, strategic bombs for the B-52 and artillery-fired tactical weapons. These kinds of modifications to existing nuclear technology — what the bomb people call "weaponization" — are the lab's bread and butter. But Lawrence Livermore Director Roger Batzel made it clear in his 1980 State of the Laboratory address that the lab's preference runs to "advanced concepts" in weaponry — the sexy, new ideas that attract the best and brightest scientists and engineers. These include the latest version of the neutron bomb and the new "reduced-residual-radioactivity" weapon, a tactical bomb that creates much less fallout but still packs a big blast.

Another bit of exotica under development at Lawrence Livermore are the "death rays" that will knock enemy missiles and satellites out of the sky. They come in several varieties.

Particle beams, very high-density and high-energy rays of sub-atomic particles, are still in the early phases of research. A 1979 report by four MIT professors argued that a particle beam system would be "extraordinarily difficult if not altogether impossible" to develop and that countermeasures would be "easy and inexpensive to deploy." But other, more eager experts have managed to secure $42 million for the construction of a controversial "Advanced Test Accelerator" at Lawrence Livermore's Site 300 that the Defense Department hopes will determine the weapon's feasibility by the late 1980s.

If particle beams are a flop, laser beams may turn out to be the military's best bet. In February 1981 Livermore successfully tested a laser powered by energy from a small nuclear explosion at the Nevada test site. The device is designed to be launched into orbit on laser battle stations that could, in the words of the Defense Department, "alter the world balance of power."

Lawrence Livermore has benefited from the energy crisis as well as the arms race. The lab gets considerable public relations mileage out of its energy research, but in reality most of this work has weapons applications. In fact, the labs are required to limit their non-weapons activities to projects that are "complementary" to their weapons work, and Director Batzel has stated that the new energy and environmental programs "are compatible with and have enhanced our work in weapons research."

Laser fusion, for example, is being touted in many quarters as the energy source of the future. It can also be used, however, to simulate the effects of nuclear weapons explosions. This may be particularly useful in the event, however unlikely, of a Comprehensive Test Ban Treaty. The lab's laser fusion work has so far been funded as a defense program costing $58 million. Another $195 million is being spent on construction of the giant Nova laser facility. It is the largest program in laser fusion in the United States.

Lawrence Livermore is the national research center for another kind of fusion, magnetic mirror fusion. In addition to a yearly $50 million operating budget (1981), the lab is constructing a $226 million Mirror Fusion Test Facility. The machine will use powerful magnetic fields to confine a super-hot hydrogen gas, called plasma, similar to the fuel that could one day power fusion reactors. Other nuclear energy work, like the testing of spent reactor fuel storage at the Nevada test site, is also part of the lab's repertoire. A relatively miniscule portion of its budget goes for fossil fuel, solar and wind research.

The Bay Area's Three Mile Island?

On the morning of January 24, 1980, an earthquake measuring 5.5 on the Richter Scale rattled the Livermore Valley, causing more than $10 million damage at the lab and forcing the evacuation of many of its 7400 employees. The quake damaged the Shiva laser facility and a small research reactor, which has remained shut ever since. All day lab officials resolutely denied that there had been any radiation releases, but that evening reporters discovered that a 30,000-gallon tank containing a mixture of water and low-level radioactive tritium had been leaking at the rate of a quart per minute. Public Information Director Jeff Garberson had been given a note about the leak in the early afternoon, but, he said, it "must have just slipped my mind."

This was one of the latest in a series of 18 accidents since 1960 which the lab admits presented potential "off-site significance." Many of these were not reported to the public at the time. The list includes two large releases of gaseous tritium (radioactive hydrogen) and smaller emissions of radioactive cobalt, zinc, iodine, strontium and plutonium into the Livermore sewer system. Twice in 1980 — within nine days — the lab accidentally leaked minute but potentially lethal amounts of plutonium, once into the atmosphere. The DOE later accused the lab of disregarding safety regulations, of inadequate management and of allowing inexperienced workers to handle badly designed equipment.

The lab routinely emits tritium and argon-41 into the air. Critics from the UC Labs Conversion Project charge that there are further, unreported releases of plutonium, curium, americium and other dangerously radioactive elements.

Whether routine or accidental, radiation escaping from the Livermore site is within easy contaminating distance of the drinking water for millions of Californians. The Hetch Hetchy Aqueduct, which supplies water to San Francisco, runs within three miles of the area. The South Bay water system passes within a few hundred feet of the lab in an open-air canal.

In 1973 the lab found plutonium in the topsoil of a field that borders the South Bay Aqueduct. The deadly material had apparently blown into the field from open-air evaporation pools full of plutonium-containing liquids. Open-air evaporation was then, belatedly, discontinued.

The danger of a massive release of radiation from the lab is enough to have prompted five Northern California congressmen to request in 1979 that

DOE remove the plutonium. Governor Brown and various local officials added their pleas, but DOE refused. Lawrence Livermore presents no "credible hazard" to people at the lab, wrote back the DOE's assistant secretary for defense programs, "much less the populations of the Bay Area."

The 1980 earthquake in Livermore, however, did little to reassure Northern Californians that they weren't living next to a potential Three Mile Island. The lab is, in fact, surrounded by 13 earthquake faults. A building that houses tritium at the Sandia Laboratory, just across the road from the main lab, sits directly on the newly discovered active Las Positas Fault. If a severe quake and fire should rupture a building containing plutonium, says Dr. John Gofman, a former associate director of the lab, it could "cause an epidemic of tens or even hundreds of thousands of lung cancers in the Bay Area." A report released by Friends of the Earth warns that a radioactive disaster at the lab could force the evacuation of many of the 4.5 million people who live within a 40-mile radius of the facility.

A controversy has also erupted over the unusually high cancer rates of employees at Lawrence Livermore. After considerable foot dragging, the California Department of Health Services released a study showing that, between 1972 and 1977, employees at the lab contracted melanoma, a form of skin cancer, five times more often than residents of surrounding communities. Twenty-seven employees are dead or suffering from the rare disease and at least two young children in the Livermore Valley have recently died from it, though elsewhere it is almost unheard of in children under 15.

Medical experts do not fully agree on the causes of melanoma. Some studies have established a link between the disease and the type of radiation emitted from materials used at the lab, but traditionally melanoma has been attributed to ultraviolet radiation from the sun. Accordingly, Public Information Director Garberson tried to blame the melanoma epidemic on the lab's "active recreation program." Director Batzel went along, issuing a memo that told lab employees, "If you are concerned about skin cancer, stay out of the sun."

Most, if not all, of the radiological hazards described here will be increased by the Reagan Administration's nuclear policies. The biggest nuclear weapons buildup in history will mean more nuclear warheads (17,000 new strategic warheads, according to the Center for Defense Information), more weapons tests (Lawrence Livermore wants to double the number of nuclear explosions at the National Test Site) and new weapons systems designed and tested at the Livermore site. The lab will be handling, storing and transporting more radioactive materials than ever before. It will also, as a result, be exposing Bay Area residents to potentially unprecedented levels of radioactivity.

RADIOACTIVE RESEARCH AND DEVELOPMENT

ELEANOR SMITH

Tucked away in an unassuming, single-story building on the peaceful University of California campus at Davis, hundreds of beagles are each day silently zapped by 60 millirems of ionizing radiation. Few of the bicyclists passing the building know about this experiment in low-level radiation, or even that there are beagles inside, because the dogs' vocal cords have been cut to prevent them from barking.

These tests, which are performed under contract for the U.S. Department of Energy (DOE) and the Nuclear Regulatory Commission (NRC), are part of a nationwide research effort by private and public laboratories into various aspects of nuclear technology. Federal agencies finance the bulk of this multi-billion-dollar industry, but state and private agencies also fund some of it. In California alone, an estimated $1 billion is spent annually on nuclear-related research.

In the case of the voiceless beagles at Davis, the researchers' purpose is to study the health effects of low-level radiation, especially bone-seeking isotopes such as strontium-90, cobalt-60 and radium-226. Other California-based research efforts are directed to medical, energy or weapons-related technologies. The operators of these facilities must, by law, report their activities to various regulatory agencies, but their record on environmental monitoring, safety procedures and security precautions is not always good — a disturbing fact, since many of these operations are located in densely populated areas.

The UCLA Reactor

A growing controversy is centered on the 18 research reactors and four large plutonium laboratories in California. At most of these facilities — universities, corporations and research institutes — relatively small but potentially lethal amounts of radioactive materials are used on a regular basis. In at least one case, the issues involved seem to be a microcosm of the problems of nuclear power: At the University of California's campus in Los Angeles, a debate is now raging over the safety of that school's small research reactor. Opponents of the plant, led by the Committee to Bridge the Gap, charge that the reactor emits hazardous levels of radioactivity, poses a real danger of a meltdown and is far too susceptible to terrorism and sabotage.

The UCLA reactor is old for a nuclear plant; it was built in 1959 by AMF Corporation, now better known for its Harley-Davidson motorcycles than its experimental nuclear reactors. Government and UC officials were surprised that anyone would bother about the seemingly innocuous plant. They contend that the reactor, which is used for commercial as well as academic purposes, is safe and poses virtually no health or safety hazards. "The maximum power output is one-thirty-thousandth of a nuclear plant," says UCLA spokesman Tom Tugend, "or enough power to operate about 100 table toasters." Adds the NRC's Jim Hanchett, "There's no way we would put a nuclear reactor in a university unless it were absolutely safe."

But critics of the plant, such as Dan Hirsch of Bridge the Gap, claim that the reactor is too old to be operated safely. Hirsch quotes from a 1976 NRC report that "some of the reactor instrumentation is still workable, but sometimes unreliable." In addition, according to an NRC inspection report, the plant may have serious problems with radioactive emissions. During normal operation, the reactor releases argon-41 in concentrations exceeding federal limits and into prevailing winds, which take the radioactive gas from the reactor stack into the air-conditioning duct of the mathematics building.

Hirsch also claims that the reactor presents a security problem. "It's operated by inexperienced college students and visitors," he says, "and, in violation of its licence, by high school students." Furthermore, Hirsch points out that the reactor uses significant amounts of weapons-grade uranium fuel. Unlike power reactors which use three-percent enriched fuel, research reactors like the one at UCLA use a 93-percent enrichment — suitable for making a nuclear bomb.

In 1981 the Committee to Bridge the Gap moved to close the UCLA reactor by challenging its relicensing before the NRC. It was the first time that the relicensing of any reactor in the country had been challenged, according to Dan Hirsch. The university, however, plans to fight for the reactor at every step of the way. It hopes to receive permission to operate the 22-year-old nuclear plant through the year 2000.

A Tale of TRIGAs

The controversy over nuclear research in California extends 100 miles south of the UCLA reactor. There, sprawled across 400 acres of prime Southern California real estate, is one of the state's most important nuclear businesses. The General Atomic Company sits in a wide expanse of mesa and valley in the exclusive San Diego suburb of La Jolla. Founded in 1955 by General Dynamics, the company is now owned by Shell Oil and Scallop Nuclear. General Atomic (GA) can boast a long history not only in "the peaceful uses of atomic energy," but in receiving huge amounts of U.S. tax dollars as well.

General Atomic spent nearly $1 billion of public and private money to develop the high-temperature gas-cooled reactor (HTGR), an uneconomic, problem-ridden power plant that the nuclear industry virtually abandoned during the mid-1970s. At one point, GA hoped to use the HTGR plan to design 25 percent of the commercial nuclear plants in the United States. Today, GA provides only fuel to the one operating HTGR in the country — the underpowered Fort St. Vrain nuclear plant in Colorado.

In 1978, General Atomic was accused by Exxon Corp. of fixing the price of uranium. GA had filed suit against Exxon Nuclear Company for $250 million in damages, accusing Exxon of breach of contract in the sale of uranium. But days later, Exxon filed a countersuit, accusing GA of participating in an international uranium cartel. The two corporations settled out of court in May 1981.

GA presently works on various high-technology energy systems, including what it calls "the largest effort in private industry toward the development of a controlled nuclear fusion power reactor. . . ." The company also manufactures radiation monitors and information systems and is a major recipient of funds from the U.S. Department of Energy.

It is the TRIGA reactor that seems to have brought GA its greatest success. The company boasts in its literature that it has sold "more than 60 of them to universities, government and industrial laboratories and medical centers on five continents." The company recently sold TRIGAs to Japan, Romania, Malaysia and Morocco and maintains an active business servicing the plants with parts and fuel.

The TRIGAs are used for training, research and isotope production. There are six of the reactors in California alone. General Atomic operates two of them at its La Jolla plant. The reactors are small; the TRIGAs at GA range from .25 megawatts to 1.5 megawatts. This compares to the 1100 megawatts of large commercial reactors like those at San Onofre.

Northrop Corporation in Hawthorne (near Los Angeles) and Aerotest in San Ramon (near San Francisco) both operate TRIGA reactors. Aerotest specializes in neutron radiography, an imaging on film of parts or materials, similar to an X-ray, but often more powerful — and dirtier. Northrop, a major defense contractor, declined to state what it uses its TRIGA for, but it is possibly used in radiography as well. Northrop manufactures fuselages for commercial and military aircraft and a wide range of electronic equipment.

Other TRIGA reactors on the University of California campuses at Berkeley and Irvine have been used by students of nuclear engineering for research and study — the TRIGAs "activate" small particles that would otherwise not be radioactive. They have helped scientists study, among other things, mineral imbalances in women taking oral contraceptives, the effects of weightlessness on bone calcium and the origin of pottery shards from Europe and the Middle East.

Noted physicist Freeman Dyson, who along with Edward Teller and Theodore Taylor conceived the TRIGA reactor, said that their goal was "to design a reactor so safe that it could be given to a bunch of high school children to play with, without any fear that they would get hurt." These scientists may be disappointed, then, with the reception the TRIGA has met in Berkeley.

Gus Newport, the mayor of Berkeley, wants the TRIGA in his town shut down. He and other local residents are concerned about the research reactor's routine releases of argon-41 and other radioactive gases. They are also worried that some of the reactor's 18 grams of plutonium could escape during a major accident or earthquake. The TRIGA sits just 100 yards from the active Hayward Fault.

Just as at UCLA, critics in Berkeley charge that security is too lax at the research reactor. The TRIGA is used in seven courses within the departments of nuclear engineering, mechanical engineering, chemistry, public health and biophysics; 100 undergraduate students and 80 graduate students attend these classes.

Dr. Joan McKenna, a local physiologist who does cancer research, says "Berkeley is not a place to have an exposed facility. Rancho Seco (a commercial reactor near Sacramento) has acres and acres around it. But anyone can walk in here. Anyone can be a student and take reactor courses. Suppose we have an act of sabotage or terrorism? Why have the risk here?"

As in the UCLA debate, however, the university believes the reactor is too small and too well designed to create any danger. Attempts by local activists and the City of Berkeley to close the reactor have been ineffective. For the foreseeable future that means anti-nuclear Berkeley will continue to have its own nuclear plant.

Three other university campuses also have research reactors: UC at Santa Barbara in residential Goleta; California State University at San Luis Obispo; and Stanford. Cal State's reactor was shut down in 1980 and Stanford's was closed in 1974.

The Big Plutonium Labs

No survey of nuclear research and development would be complete without the inclusion of four facilities well known to those who have watched the growth of atomic energy in California: Atomics International, Lawrence Berkeley Laboratory, Lawrence Livermore Laboratory and Vallecitos Nuclear Center. Each of these facilities operates a "hot" laboratory for handling plutonium. Workers stick their hands into a protected "glove box" which stretches into a sealed room. Inside they perform experiments with pluto-nium and a potent handful of other radioactive elements. Alongside the plutonium labs at Atomics International, Lawrence Livermore and Vallecitos are California's eight remaining research reactors, bringing the state's total to 18.

The Lawrence Berkeley Laboratory is careful to distinguish itself from its namesake at Livermore. "Lawrence Berkeley Lab conducts fundamental re-search on the nucleus of atoms" and "a great deal of medical research," explains Public Information Director John Feack. The latter includes radio-therapy techniques on human cancer patients and tests on mice to remove or "sequester" plutonium from the body.

The lab uses accelerators, but no reactors, in its research into nuclear physics. Accelerators are used to study subatomic particles by hurling and smashing atoms at high velocities. It was, in fact, at the labs which even-tually became LBL that Ernest Lawrence invented the cyclotron — a kind of accelerator — for which he won the Nobel prize in 1939. This work was followed one year later with the secret discovery of plutonium by a team of chemists at the labs and led to an effort which eventually produced the world's first atomic bombs — the Manhattan Project.

Today, according to LBL's Feack, the lab does "no research on nuclear energy or nuclear weapons." In its work into nuclear medicine and physics, however, it does handle up to 500 grams of plutonium and generates a moderate amount of radioactive wastes. All this occurs in residential Berk-eley, just up the hill from the University of California and the powerful Franklin earthquake fault.

About 20 miles southeast of LBL, nestled in the foothills between Sunol and Pleasanton, General Electric's Vallecitos Nuclear Center conducts wide-ranging research into nuclear fuels and radioisotopes. For this, the facility operates two research reactors, the GETR (GE Test Reactor) and the NTR (Nuclear Test Reactor). The NTR generates only .1 megawatt, but the GETR is built for 50 megawatts, nearly the size of PG&E's commercial nuclear plant at Humboldt Bay. Two other reactors at the site are being decommissioned.

Also at Vallecitos is a facility which GE calls its Plutonium Fuels Labora-tories. Here, company technicians and scientists take apart spent fuel rods, test plutonium reprocessing methods and conduct research into fueling breeder reactors. GE is licensed by the NRC to handle up to 150 kilograms of plutonium at the plant, an amount which has stirred controversy among some residents of the Bay Area.

In October 1977, after challenges from environmental groups, the NRC ordered the GETR shut down, pending a seismic safety review. However, it did allow the plutonium labs to continue operating. One year later, samples of drinking water were taken from wells downstream from Vallecitos which contained levels of radioactive tritium two-and-a-half times the federal limit.

Tritium in drinking water has been linked to ovary damage and cancer in mice. But the Environmental Protection Agency, which did the tests, did not close the wells. EPA officials pointed out that the standard refers to average annual intake; thus, they concluded the wells were safe. No one knows, however, how much tritium or other radioactive isotopes may have leached into the drinking water over the last 20 years. GE was on a self-monitoring program, during which time no government inspections of air or water were made.

The two other large plutonium laboratories in California have also raised the ire of the state's nuclear critics. At the federal government's Lawrence Livermore Laboratory east of San Francisco, work revolves around designing the most advanced nuclear weapons in the world (see page 95). At Atomics International in the San Fernando Valley near Los Angeles, employees of Rockwell International conduct research and development into both nuclear energy and nuclear weapons for the Department of Energy (see page 24).

Taking Out the Garbage

All of the nuclear research centers in California — the universities, the corporations and the federal labs — discharge significant amounts of radioactive waste. Most research reactors, unlike the big commercial power plants, use highly enriched fuel which burns slowly and generally lasts the lifetime of the reactor. The high-level radioactive waste, in the form of spent fuel, is removed only at the end of the reactor's useful life. It is either stored on site, sent off site for reprocessing or shipped to DOE-owned facilities in Richland, Washington, or the National Test Site in Nevada for permanent storage.

California's research facilities also make frequent shipments of less contaminated materials, such as radioactive clothing and tools. The ultimate resting place for this low-level waste, including the unfortunate irradiated beagles from UC Davis' Radiobiology Laboratory, is the commercial radioactive waste dump at Beatty, Nevada.

Michael Gillogly

PART SIX

CLEANING UP THE NUCLEAR STATE

Like a bad dream, California's radioactive wastes have come back to haunt us. We now discover that thousands of barrels of the wastes were dumped in heavily fished waters off the California coast; that the state's reactors might close because there is no safe place to store their lethal by-products; and that eventually even the reactors themselves must be painstakingly wrenched apart, shielded from the public and shipped out of state.

These are some of the stubborn problems presented by the next four articles. The authors leave us facing the dilemma of how to clean up a state now littered with radioactive wastes. It is the final bill left by Nuclear California, and it must be paid for soon.

Photo: The reactor and twin cooling towers of the Rancho Seco nuclear power plant

BURIAL AT SEA

DOUGLAS FOSTER

A blue fishing boat called the *Marilyn J* sliced through ocean swells the color of dishwater while the sun dropped through a pink flush behind the Golden Gate Bridge. Men hoisted their catches — black cod, butterfish — above their heads, gripping the fish in oily plastic bags and waving them toward San Francisco civic leaders and state officials waiting on the dock. This was no ordinary fishing trip. It was a periodic check, at taxpayers' expense, to determine whether a prime fishing area which supplies the city's Fisherman's Wharf with butterfish, squid and crab had been contaminated by radioactive trash.

It is not generally known that during the past 35 years the oceans have been used internationally as junkyards for more than 76,000 tons of nuclear garbage. In the United States alone, 50 dumps were sprinkled along both coasts. Nearly all of them are located close to densely-populated cities, in prime fishing grounds within a few hours' boat ride from New York, Newark, Boston, Los Angeles and San Francisco.

San Francisco's dump, a 600-square-mile expanse of sea surrounding the Farallon Islands, is the most famous of the radioactive junkyards established by the federal government in 1946, and then used until 1970. The site is the largest known dump off American shores and has achieved fame partly because public concern in Northern California has risen from benign neglect to forceful tides of fear and outrage.

Spokesmen for federal agencies overseeing the dumps — retired brass from the Atomic Energy Commission (AEC), and present staff at the Environmental Protection Agency (EPA) — have steadfastly insisted that there is "no evidence of any harm to either man or the marine environment" from jettisoning radioactive trash into the ocean near San Francisco or elsewhere. But while over $20 million has already been spent to find ways of stashing high-level nuclear waste *beneath* the ocean floor, a mere $250,000 has been spent to assess the damage from the *existing* dumps. Scientists who have reviewed the monitoring program believe the federal government's efforts have been flawed from the outset, revealing little about the existing environmental dangers from the dumps. Several critics of the federal monitoring — including Representative Glenn Anderson (D-Long Beach), Representative

John Burton (D-San Francisco) and Governor Jerry Brown — doubt that federal officials have any convincing data to support their reassurances of safety.

There is evidence to show that federal officials have little idea how much radioactive garbage lies off each coast, where the radioactive canisters which hold it are and even whether large numbers of fish have already been contaminated.

At congressional hearings and press conferences, the EPA trots out a set of figures which constitutes the government's best guess of the inventory of nuclear garbage dumped into the ocean. Government officials believe that 47,500 barrels filled with radioactive trash were dumped at the Farallons, containing 14,500 curies of radioactivity. (A curie is a measurement of radioactivity equivalent to that emitted by a gram of radium.)

Although five other dump sites were officially established off the California coast, EPA oceanographer Bob Dyer — who has been in charge of all dump monitoring — said that more than 90 percent of the refuse dropped into California's coastal waters was sunk off San Francisco. "As far as trusting the records, they are the only records available, and I have no reason to suspect that I shouldn't trust the records because the sea disposal that was occurring at the time was done with public acceptance. They were reporting it as it was; that's my opinion," Dyer said.

Despite Dyer's trust in the old AEC records he inherited, even a cursory perusal of the data reveals that either federal officials are making fanciful guesses or that a boggling form of New Math is employed. Watch carefully now as the statistics get jumbled.

In a 1957 report, AEC researcher Arnold Joseph wrote that through 1956, 12,500 barrels had been dumped in the Pacific Ocean off San Francisco, garbage with a radioactive count of 10,000 curies. In 1960, the AEC told Congress that 21,000 barrels had been dumped, with 14,000 curies radioactivity. In 1980, EPA comes up with its seminal estimate: 47,500 barrels with a 14,500-curie content. From 1956 to 1970, when the dumping stopped, the number of radioactive canisters went up 400 percent, but the radioactive inventory increased only 26 percent, if government data is to be believed. Joseph himself was more careful, noting that the radioactive inventory contained in his report could be off by as much as "a factor of 10."

Leave aside, for a moment, that particular numerical jungle. There are other pieces of the puzzle, enough to riddle the official estimate with doubts:

• A man who packaged waste for Lawrence Livermore Laboratory in 1954 remembers averaging 40 barrels *a week*. At that rate, Livermore alone could have generated more than 30,000 barrels, and there were two other laboratories in the San Francisco Bay Area generating wastes at the time and five licensed disposal firms in California.

- Military dumpers jettisoned radioactive cargo wherever it was convenient, ignoring AEC regulations and apparently keeping no records of how much nuclear waste was dumped. Retired officers have told Congressman John Burton of San Francisco that the military routinely air-dropped nuclear waste into the Pacific Ocean. "Once a week, from 1952 to 1967, the U.S. Air Force stationed at Hamilton Air Force Base dropped nuclear waste approximately one-half mile from the Farallon Islands," Burton claimed.
- No land disposal dump sites were established until the early 1960s. With the military producing more than 90 percent of all nuclear waste, it is legitimate to ask where they *did* place nuclear garbage if not in the ocean.

Perhaps even more bothersome than the questions about the quantity of radioactive trash dropped into the sea are the recurrent indications of very hot material being dumped. Federal officials insist that only low-level waste was accepted for ocean disposal, mostly "paper towels, rags, broken glassware, clothing and other laboratory paraphernalia contaminated with trace amounts of radioactive materials."

But Congressman Glenn Anderson has been haunting federal officials for more than a year about the admission of a former university researcher who acknowledged that "high-level liquid waste" had been dumped in the ocean. Navy documents only recently released show that, despite Department of Defense denials, the military dropped nuclear waste off Los Angeles on a regular basis. Those documents contain vague references to reactor fuel elements and samples, references which have yet to be explained by Navy officers.

Long-buried Atomic Energy Commission records demonstrate that the government was aware that firms licensed to dump radioactive waste took shortcuts, accepted high-level material illegally and kept scanty records. In one case, Coastwise Disposal Company of Long Beach was accused of hauling waste away from an Atomics International plant near Ventura, where a nuclear meltdown took place in late 1959, without asking what the barrels contained. The firm's license was yanked after an explosion at its dock in late 1960, but not before it dumped 650 tons of radioactive garbage off the coast of Long Beach.

On May 13, 1958, California state officials and representatives of the firms licensed to drop radioactive waste off the California coast met in Berkeley to hammer out an agreement about where, and at what depths, nuclear trash could be dumped. They agreed on the Farallon Islands site, Port Hueneme, Los Angeles, San Diego and the Santa Cruz basin. But the provisions for dumping insisted on "a safe means of disposal which will adequately protect human and marine life." The representatives agreed all wastes would be jettisoned in "at least 2000 fathoms" (12,000 feet) in "containers of such integrity that they will remain intact at the depths prescribed."

Nearly two decades later a group of deep-sea divers led by Dr. Harold Ross of Project Tek-Tite, a private organization of marine enthusiasts, was searching for a piece of lost equipment just 18 miles from San Francisco. Diving in less than 160 feet of water between the Farallon Islands and the Golden Gate Bridge, one of the divers saw something that brought him up in a hurry.

"The guy came up all flushed and he said, 'Doc, there's barrels down there'. . . . When you've got a young crew and you talk about barrels, the first thing they think about is treasure. So we all went in the water," Ross remembers.

The Tek-Tite divers were disappointed, and frightened, when they discovered there was no treasure at the bottom. Several dozen barrels had been corroded away, and the contents were spilling out. Large pieces of laboratory equipment were visible, material was scattered along the ocean floor and there were fish feeding nearby.

The AEC had required only a depth of 6000 feet for the disposal of radioactive garbage, and the licensed companies had simply ignored their agreement with California officials. But even the AEC restriction turned out to be a paper protection. None of the dump sites was surveyed to ensure that waste would land in areas of sufficient depth, and barge operators were given large latitude in determining where to drop their nuclear cargo.

The dangers of radioactive leaks beneath the sea would likely have remained a buried issue in the 1980s if not for the persistence of environmental groups and the considerable energy of a soft-spoken, unassuming scientist from the University of California. Jackson Davis, a marine biologist at UC's Santa Cruz campus, was launched onto center stage of the ocean dumping debate when he teamed up with San Francisco Supervisor Quentin Kopp and Dan Hirsch, a spokesman for the Los Angeles–based Committee to Bridge the Gap, to demand the release of government information about the dumps.

In July 1980, Hirsch had uncovered evidence of 50 nuclear waste dumps spread down both coasts like a bad case of poison oak. At the time, federal officials were acknowledging only four dumps. Both Hirsch and Davis were also concerned about the EPA's secretive approach. Surveys of the existing dumps which had been conducted in 1974 and 1977 had been kept under tight wraps, while summaries prepared by Dyer assured there was "no evidence of any harm to either man or the marine environment."

Once the EPA survey data had been dislodged, Davis charged the information "furnished compelling evidence that radioactive contamination from the dump sites has entered edible fish and now presents a measurable health hazard." At the Farallon Islands dump, radioactive levels in bottom sediments were 2000 times what scientists expected from "background" — a level combining natural radioactivity and the plutonium added to the environment by years of atmospheric testing of atomic weapons. Off New Jersey, the readings were even higher: up to 260,000 times higher than the level expected from fallout.

Perhaps most important, Davis found that a colossal error had been made in government plans. From the inception of ocean dumping in 1946, AEC officials pursued two different strategies for dealing with nuclear waste: one, release the toxic elements into the air or water and dilute them to safe levels; or two, contain and store the most dangerous waste until the radioactive readings fall to harmless levels. Davis could see that the government strategy had been the worst possible merger of the two methods.

AEC officials had expected their garbage to remain intact until it hit the ocean floor, there to leach bit by bit into sea water and disperse. But by thoroughly reviewing government documents, Davis found that nearly a third of the radioactive canisters crumpled on their way to sea bottom. Instead of diluting, radioactive isotopes imbedded themselves in ocean mud, sticking there in concentrated form. Perhaps most disturbing, the radioactive canisters created an artificial environment attractive to marine life — a kind of condominium project for soft fish wanting to get out of the currents. Sponges attached themselves to the barrels, small worms ate the leaching waste material and spread it into the bottom sludge. Fish could eat the worms, of course, concentrating the radioactivity in their systems many thousands of times.

"You couldn't actually design a better way to put that radioactivity in our food," Davis said.

Although state researchers have not found fish with astronomical radio-activity readings near the Farallons, only a small sample has been taken so far and the findings have been "luck, not planning," according to one top EPA official.

The story of radioactive ocean dumps is a catalog of ill-conceived, uncontrolled and secretive government operations. At every step in the history of ocean dumping, government scientists were profoundly ignorant of the consequences of their acts. When the dumping began, AEC analysts were certain that food fish never ventured more than 400 fathoms (2400 feet) below the surface. In 1962, the National Academy of Sciences found commercial fish at 1200 fathoms, *1200 feet deeper than the legal restriction for nuclear waste.*

In 1962, the Academy also recommended that radioactive ocean dumps be "exhaustively surveyed and sampled for organisms living on or near the ocean bottom" and urged routine monitoring of the dump sites. Instead, U.S. officials waited four years after the dumping had stopped — then took a few photographs and samples on a shoestring budget of $250,000 before pronouncing the dumps safe. These pronouncements harken back to the repeated assurances of the AEC in late 1961. "Report Shows No Radioactivity Attributable to Waste in Two Pacific Sites," one press missive was headlined, above an ironclad finding that all waste was "safely contained" 6000 feet below the surface at ocean bottom. The early reports were met by a trusting public. The more recent EPA reports have encountered skepticism.

Repeated entreaties to conduct systematic studies of the dump sites have been rebuffed, and legislation has been introduced to force the government's hand and require extensive monitoring. Even Dr. William Schell, a radiological specialist who compiled much of the EPA's sketchy data on fish, has been appalled by the government's strenuous efforts to remain ignorant about the dumps. While more than $20 million is spent to design ways to jettison high-level nuclear waste onto the ocean floor, nothing is budgeted for an assessment of health hazards from material already dropped into the sea. "What we don't understand is the transfer vehicle, how the radioactivity gets through the food chain to man. . . . It's surprising to me that nobody will fund such a study," Schell said.

What do the government officials, who repeatedly assure us that ocean dumping poses no hazard, really know? Not how much radioactive garbage lies off each coast. Not where it is, at what depth, in what condition. Not whether large numbers of fish have been contaminated already. Government officials cannot even be sure when radioactive canisters can be expected to release their toxic contents.

"If you assume that an undamaged barrel will rot in 40 years and a damaged barrel would rot in 20 — which is my best assessment — I would imagine that you would see a steady buildup of activity . . ." Davis said recently. "I'd say peak release would occur in the 1980s and 1990s.

"Of course, the full impact of the health hazard depends upon the exact composition of the waste, something we may never know. But if it is plutonium, then the toxicity duration is the equivalent of 20 half-lives, which is 500,000 years. So that's the period of time we're talking about, 10,000 human generations — which is not our grandkids, not their grandkids, but whole new civilizations somewhere downstream."

DOIN' THE NUCLEAR WASH

MICHAEL KEPP

Since 1959 the hottest laundry in California has been operating on a residential block in Pleasanton, a small, thriving city in the Livermore Valley southeast of Oakland. Along with the grease, grime and ground-in dirt, the clothes washed are full of plutonium, uranium and a dozen other kinds of radioactive waste. A Geiger counter — not a ring-free collar — is the gauge of whether or not the clothes are clean.

California's only off-site nuclear laundry, Interstate Nuclear, doesn't consider itself a laundry; it's in the "decontamination" business. Interstate picks up irradiated clothes at nearby nuclear reactors and radiation labs. The company then takes everything from coveralls and cloth hoods to rubber gloves and respirators and washes them in 500-pound-capacity stainless steel behemoths that dwarf your average Hotpoint.

It's an endless grind at Interstate. Clean clothes last only a few hours at the workplace before they're dirty again. But the customers don't seem to complain. Why should they? Lawrence Livermore Laboratory and Vallecitos Nuclear Center can hardly take their business to Eddie Yee's Martinizing at the shopping center. Many newer nuclear installations have washers and driers built in, but the folks in the Livermore Valley still prefer to send out.

This practice doesn't sit too well with the people down the street from Interstate. Some worry about the trace amounts of radioactivity the laundry's ventilation stacks release. The California Health Department considers this amount below state and federal limits, but for many in the area the laundry is too close for comfort. In February 1980 Interstate applied for a permit to expand, and a group of neighbors decided to challenge the appropriateness of such a business on a residential block. They campaigned at city council hearings, and their efforts denied the expansion. The city now wants to force Interstate to clean its clothes elsewhere.

Interstate is a little hot under the collar. The $60 million, Massachusetts-based outfit operates similar laundromats in six other towns from New Kensington, Pennsylvania, to Bremerton, Washington. But the Pleasanton plant is a big money-maker, and Interstate would like to continue laundering without being hounded by an irate citizenry.

Interstate manager Craig Connelly called the campaign a "witch-hunt" and swore to the benign nature of his supercharged Speed Queens. "I've been in this business for 11 years," said Connelly, "and I don't glow."

THE RADIOACTIVE WASTE DILEMMA

MARK VANDERVELDEN

The stubborn technical and political stalemate over the disposal of nuclear waste has become a scientific Vietnam for the nuclear industry. Fully four decades since the first lethal high-level and low-level nuclear wastes began accumulating, fundamental scientific issues remain unresolved, if not unresolvable. Equally important, countless institutional, social and political obstacles stand in the way of reaching consensus on the growing predicament of waste disposal. All the nation has to show for its efforts to safely manage nuclear waste is a litany of pious technological exhortations, a record of ambitious but premature false starts and a chronic history of federal and industrial mismanagement.

The failure to develop an acceptable final solution to the problem of nuclear waste has, in California at least, cast the nuclear industry into increasing political turmoil. With the outlook for a permanent waste disposal operation before the turn of the century looking bleaker, California utilities face the sobering possibility that existing reactors could be forced to shut down because there simply is no safe place to store their highly radioactive spent fuel. Of far greater concern to the public, however, is the grim prospect that California, by default, could wind up as the eternal babysitter for thousands of tons of deadly long-lived waste for which nobody bargained.

California's Nuclear Waste Dilemma

California's nuclear waste disposal problem is serious, and getting worse with each passing year. The nuclear industry proceeded on the assumption that spent fuel from commercial reactors would be reprocessed to reclaim usable uranium and plutonium. The remaining wastes, it was believed, would not pose any insurmountable technical difficulties. But the industry missed the mark badly. Every effort made so far to make reprocessing economically and technically feasible has been a dismal failure. Complicating matters, President Jimmy Carter announced a new policy in 1977 which has had the effect of putting domestic commercial reprocessing into indefinite limbo (although the Reagan Administration now hopes to revive it). Reprocessing, Carter argued, held the threat of global spread of nuclear weapons.

The net result has been a buildup of spent-fuel assemblies at reactor sites which were never designed for long-term storage. Utilities in California and elsewhere have been forced to store their spent fuel on site within the radio-active confines of specially designed pools of water.

The Problems with On-Site Storage

Although the nuclear industry has had a good record of success with storing spent fuel on reactor sites, there remains a number of pressing safety concerns. Reactor storage pools were never designed to store spent fuel for extended periods of time. There is no experience with spent-fuel storage over a 20-to-50-year time span. Little is known about the effects of corrosion on spent-fuel rods stored underwater for long periods of time. If utilities decide to expand their storage capacity by re-racking the inside of the pools, new concerns will arise over the safety of so much intense radiation within such a small place. Critics fear that if the fuel rods are spaced too closely together, an accident similar to a reactor core meltdown could occur. In addition, there are security considerations. With so much dangerous spent fuel concentrated in one spot, a storage pool could become a tempting target for sabotage or terrorism.

The overriding question here is whether or not "temporary" on-site facilities will, by default, end up as permanent or nearly permanent nuclear waste dumps. Industry and government officials view that prospect as absurd. Nevertheless, with fuel reprocessing on the economic and technical ropes, and with no federal storage bailout on the near horizon, the remaining alternative appears to be permanent geologic disposal. Like other options, geologic disposal has plenty of problems.

"Let's Just Bury the Damn Stuff . . ."

America's waste culture mentality ("out of sight, out of mind") best characterizes the intuitive appeal of burying nuclear waste deep in the earth. Over the years a variety of schemes have received serious consideration as permanent waste disposal strategies. These suggestions have ranged from rocketing the waste into the far reaches of space to burying it in the Antarctic ice cap or ocean floor. None of these ideas has gotten past the drawing board stage. Most experts are pinning their hopes on the concept of deep geological burial.

The idea of placing wastes in stable geologic formations seems to have originated with a 1957 National Academy of Sciences report which described geologic disposal as "the possibility promising the most immediate solution" to the waste disposal problem. For more than 10 years the NAS report stood as the conventional wisdom on the subject. But during the 1960s and early

'70s the federal government set out to prove the feasibility of geologic disposal; each effort met with failure. By 1976, it was becoming clear that permanent isolation of nuclear wastes from the biosphere would require more than the mere mining of a cavity in a salt dome and then putting waste canisters in the hole. At this point, the waste disposal issue started to become a matter of broad public concern. Virtually overnight, a highly technical debate, once limited to scientists and engineers, turned political. As the debate over what to do with the nuclear industry's lethal detritus raged in statehouses and ballot boxes, public attention began to focus on the entire question of the risks to society posed by nuclear power. It was the beginning of the nuclear industry's version of a political Vietnam.

The California Nuclear Waste Laws

By early 1976, a citizens' campaign to stop nuclear power in California was well under way. The rancorous debate on the streets moved to the legislature, where, after exhaustive hearings, three nuclear safeguard bills were passed and signed into law by Governor Jerry Brown. The three nuclear bills were widely viewed as a compromise to the tougher provisions proposed by Proposition 15. Although voters rejected Prop. 15 at the polls, it did succeed in forcing the political debate to a head and was, in retrospect, a successful effort. The California Nuclear Safeguards Act essentially prohibits the siting of new nuclear power plants in the state until such time as a permanent means of disposal of high-level radioactive waste is demonstrated. The key word is *demonstrated*. The legislature had skeptically shifted the burden of proof onto the nuclear industry.

Though the implications of this policy were not well understood at the time, it proved to be an idea which won quick acceptance around the country. In 1975, only two of the states considering nuclear statutes enacted them; by 1976, however, 16 states had laws dealing with the waste issue and another eight had laws on increased radiation controls. By May 1978, 33 of the 50 states had laws aimed at some aspect of radioactive-waste control. The Three Mile Island accident sparked yet another wave of legislation pointed at increasing state control over nuclear power plant decisions. By October 1979 some 19 states had enacted bans or moratoria specifically on the siting of nuclear waste dump sites.

Under the terms of the California statutes, the state's Energy Commission is required to report to the legislature the progress of federal waste disposal efforts. In January 1978 the Commission issued its first assessment of the problem and concluded that so many unresolved technical questions remained that it is "... questionable to assume that any waste disposal technology will be demonstrated in the near future." More recently, Energy Commission officials testified before Nuclear Regulatory Commission hearings and stated:

In no case is there cause for confidence that the National Waste Terminal Storage program will provide safe geologic disposal when needed. From a scientific standpoint, the NWTS program is in its infancy. Significant gaps exist in the scientific data which are necessary to assure that a geologic repository will isolate radioactive wastes from the biosphere.

The Energy Commission staff went on to say that the federal waste management program is in complete disarray and is operating without coordinated schedules, policies or sound management by the Department of Energy. The entire federal effort lacks credibility, the staff charged, and added that "no plans currently exist with a reasonable prospect of ameliorating these inadequacies."

The Timetable for High-Level Waste

What are the prospects that science will be able to turn the situation around? A look at what's been promised so far offers a hint. In October 1976 a statement from the White House set 1985 as the date for a licensed waste dump site. The 1985 date was still the target in 1978 when the Energy Commission released its first report on the problem. Sixteen months later, federal officials moved the date from 1985 to 1988. Eight months after that, the date was changed again to between 1988 and 1992. In President Jimmy Carter's 1980 message on nuclear waste disposal, "about 1995" became the new target. *Nucleonics Week* reported in its May 8, 1980, edition the latest Department of Energy target date as 1997.

Within three years — the same three years in which federal programs in nuclear waste disposal dramatically expanded — DOE's pronounced target date had slipped 12 years. In 1976 the target date was 1985 — nine years away. Today the goal is 15 years away — almost double the original estimates. The Energy Commission is quick to note that 15 years is beyond the horizon of credible government planning efforts, and in its estimate no waste disposal site will likely be in operation before the early part of the 21st century, if ever.

The consequences for the nuclear industry are harrowing. At the very least it would mean, under existing California law, that no new nuclear power plants could be built. Other factors, principally lower demand and economics, are likely to have the same effect on new construction, regardless of the nuclear laws. Still, with respect to existing reactors, on-site spent-fuel storage space is likely to fill up well before any disposal site can be demonstrated to work. Should the state's utilities opt for expanding on-site "temporary" storage they will undoubtedly be forced to confront strong political opposition from those who fear "temporary" could wind up being "permanent."

The best bet for the nuclear industry is to hope for yet another massive federal bailout in the form of an away-from-reactor storage facility. It's a good bet for the industry — every other time they have wagered on the federal government to get nuclear energy out of a political pinch the industry has cashed in big with huge subsidies. But if Uncle Sam does not come through, California will face a mounting crisis caused by the most lethal waste from the entire nuclear fuel cycle. A quick survey of the state's commercial reactors gives an idea of how quickly these unplanned nuclear waste dumps may appear.

Rancho Seco: California's most serious spent-fuel storage problem is probably at the Sacramento Municipal Utility District's Rancho Seco reactor. Unable to find a facility for commercial reprocessing, SMUD officials decided in 1975 to rebuild the plant's spent-fuel racks in their storage pool in order to handle more capacity. To avoid the environmental impact reporting requirements of the California Environmental Quality Act, the SMUD board of directors declared an emergency. New racks were installed, found defective and reinstalled. With no plans to ship its spent fuel off site, SMUD may run out of room by 1985, depending upon the performance of the plant. If that occurs, Rancho Seco may become the first commercial reactor in California to be shut down because there is no place to store its nuclear waste.

Humboldt Bay: Pacific Gas & Electric's small Humboldt Bay reactor has been permanently closed due to seismic hazards. More than 31 metric tons of spent fuel now sit on the site, and it's unclear just what will be done with it. Without reprocessing or any means to permanently dispose of the fuel assemblies, PG&E's best hope is for the federal government to establish a national or regional storage facility. Otherwise, Humboldt Bay may inadvertently end up as California's first long-term waste storage site.

San Onofre Unit One: San Onofre's luck held out until 1980. Its owners, Southern California Edison and San Diego Gas & Electric, had a contract with General Electric which allowed the utility to ship and store spent fuel at GE's Morris, Illinois, facility. The residents of Illinois, however, apparently grew tired of having their state used as a waste dump, and in 1980 they pressured the state legislature to ban nuclear waste shipments from states without reciprocal agreements. In other words, because California has no high-level waste storage sites of its own, it may no longer use those in Illinois. The law is now being challenged in court as unconstitutional. Meanwhile, spent fuel is piling up at San Onofre. The plant discharges 52 spent-fuel assemblies every 18 months. If the Illinois law is upheld, Edison will have to expand the plant's spent-fuel pool, make use of space from two additional reactors now under construction or shut down its first reactor.

San Onofre Units Two & Three and Diablo Units One & Two: California's four newest commercial reactors are predicted to produce approximately 25 metric tons of spent fuel per plant. SCE and PG&E have installed high-density spent-fuel racks at their plants which should carry them through the early 1990s. Beyond that point, however, these reactors will face the same problems as Rancho Seco and could be forced to close.

Low-Level Waste in California

The problems faced by California in disposing of spent fuel are similar to those posed by the disposal of low-level nuclear wastes. From a technical standpoint, low-level wastes are generally less concentrated and shorter-lived than high-level wastes. That is not to say that low-level wastes present any less of a hazard to public health and the environment. In fact, low-level wastes have often been mishandled and improperly disposed of.

Old radioactive dump sites have been discovered nationwide, including areas near downtown Denver and Salt Lake City. Much of the waste was dumped by commercial operations earlier in the century, such as the companies that produced radium for watch dials. Other dump sites consist of radioisotopes used in the medical industry, including one created by the Veterans Administration hospital in Brentwood, a community in Los Angeles. Plans to turn the site into a city park were halted in 1981 when the Los Angeles–based Committee to Bridge the Gap disclosed that the V.A. dumped an unknown amount of radioactive waste during the 1960s.

Aside from identifying and cleaning up these sites, the major problem in California is related to the sheer volume of low-level wastes produced. Each year anywhere from 150,000 to 200,000 cubic meters of low-level wastes are created by commercial nuclear reactors, medical research and other industrial processes. This would fill an area the size of a football field to a height of 100 feet. Other than New York, California is the single largest producer of low-level radioactive waste in the nation. Its 2000 state-licensed medical, industrial and educational users, two commercial power plants, 13 research reactors, nuclear Navy vessels and Department of Energy contractors produce more low-level radioactive waste than the rest of the Western states combined.

California so far has been able to unload its radioactive garbage on Nevada and Washington. The Beatty site, located about 70 miles north of Las Vegas, has been the recipient of nearly 95 percent of all California-produced low-level waste. The federal nuclear reservation at Hanford, Washington, has received the rest. The Beatty site has been a source of controversy for many years. Numerous accidents in transporting and burying the waste have resulted in a number of temporary closures. But despite the persistent efforts of Nevada state officials, including Governor Robert List, the Beatty site remains open. In November 1980 voters in Washington state elected to ban from burial all out-of-state, non-medical low-level waste effective July 1, 1981. The constitutionality of the initiative is still a matter of dispute.

With the threat that the Beatty site may be closed on a moment's notice, and the clear signal from Washington that California's nuclear wastes are not welcome, users of radioactive materials are facing a dilemma. All signs indicate that before long California will be forced by other western states to bear its burden of the low-level waste problem. Simply put, California cannot much longer avoid opening a dump site within its own borders. When and where that is likely to happen is still far from resolved. One thing is clear, however. The state's large number of radioactive users are growing anxious at the prospect of having no place to dispose of radioactive waste. Western states, meanwhile, are no longer willing to passively resist California's pressure. It is a very touchy political problem which in all likelihood will not fully emerge until the hard work of siting a dump begins. At that point, there could be intense political opposition which once again could change the face of nuclear politics in California.

The Problem That Won't Go Away

The problem of nuclear waste lies at the heart of the nuclear issue. It is virtually unsolvable by the means available to us today; it should have been addressed before the first commercial reactor was built, not when the mounting pile of wastes threatens to engulf us. Both the government and the industry have failed to devise any workable scheme to dispose of radioactive by-products. The result of this failure has been a virtual ban on new nuclear power plant construction, a severe loss of public credibility and the prospect that existing power plants may be shut down within the next decade. It is entirely possible that there is no safe way to dispose of the toxins created by the nuclear production of power. If this is the case, the blame ought to be left squarely where it belongs — on the shoulders of a nuclear establishment whose technological optimism and corporate collusion have left a radioactive legacy that will endure for generations.

WHAT TO DO WITH A DEAD NUKE

JOHN ROSS

The redwood forests of California's north coast conjure up images of our tallest trees, strapping lumberjacks, Bigfoot and the last pristine light in the nation. But behind the Redwood Curtain — a sparsely populated, predominantly federally owned area the size of your average eastern state — environmental yahooism has been the rule for decades. Boom-and-bust tractor-logging has left only a cosmetic facade of first growth along the highways. Salmon fishing is diminished in the silted rivers, and, as you read this, Weyerhaeuser poises on the brink of ecological gamesmanship with a million-dollar ocean-ranching scheme waiting in the wings. Back in the hills, the U.S. Forest Service blissfully wipes out hardwood with phenoxy herbicides.

As long ago as 1958, the north coast was selected as the site of California's very first commercial nuclear reactor. "On a barren stretch of coast at Humboldt Bay, 280 miles north of San Francisco, surveying teams this week went to work on a radically new type of power plant," reported *Time* magazine in November 1960. The first yards of concrete at Humboldt Bay were poured amid ominous synchronicity. In January of 1961 — coincidental with the inauguration of John Kennedy's New Frontier and the first American civilian nuclear disaster, at an Atomic Energy Commission test reactor in Idaho Falls — Norman Sutherland, then president of Pacific Gas & Electric (PG&E), Humboldt Bay's owner, was on site touting the new technology: "No one has yet been able to make electricity with atomic fuel anywhere in the United States as cheaply as it could be done with [other fuels]. We expect this will be accomplished here at Humboldt Bay. And you will have a new tourist attraction for visitors to this section of the Redwood Empire . . . and a laboratory for your schoolchildren." Sutherland was almost on target. For the first 14 months of the nuke's operation, the elementary school directly downwind from the stack kept film badges in the building to measure radiation.

Humboldt Bay's construction occasioned about as much controversy in the area as a new bowling alley. PG&E flacks orchestrated an extravaganza when the plant was dedicated in September 1963. Six hundred guests gathered inside a circus tent the utility had rented for the occasion. Preachers and historians were on hand for the sing-a-long, and the inaugural handouts glowed of more innocent times: "Atomic energy has a vital role to play in everybody's future."

Two months later, Humboldt Bay underwent its first two "scrams" — sudden emergency shutdowns. There would be many more.

On July 23, 1976, on the downslope of the Bicentennial, Humboldt Bay closed down for refueling — and has not reopened since. It was closed because of government regulators' rising fears about imminent ground movement. The Humboldt plant is located in an earthquake zone.

Today, this tiny facility is a leading candidate to become the largest light-water commercial reactor in the nation ever to be decommissioned. (Most previously decommissioned reactors have been government-sponsored experimental projects.) Humboldt Bay — California's first reactor — is old and outmoded and has not produced a crackle of electricity in more than four years. But Pacific Gas & Electric remains reluctant to publicly discuss decommissioning it. It appears that PG&E just does not know how to decommission Humboldt Bay. Like most other utilities in the nation, it has never before had to deal with a dead nuke.

When researchers for the anti-nuclear Redwood Alliance tried to pinpoint PG&E's decommissioning plans, they discovered that the Nuclear Regulatory Commission (NRC) does not require such plans at the time of licensing. A utility only need show that it can meet the estimated costs of "permanently shutting down the facility and maintaining it in a safe condition."

"It is assumed that if an applicant for a reactor license is financially qualified to construct a nuclear facility, it's also qualified to shut it down," explains one NRC official.

All this sounds marvelously trusting, perhaps too much so. The federal government has authorized several studies on decommissioning, but no detailed regulations exist at present.

In fact, the word *decommissioning* itself is really only a bureaucratic euphemism, which makes the process sound as facile as mothballing a World War II battleship. In reality, the dimensions of what must be done are stupefying. All 72 licensed commercial reactors in the U.S. today will have to be decommissioned in our lifetime, because reactors have a life span of 30 to 40 years. No one — the industry, its opponents or its regulators — disputes this. By the turn of the century, the Pandora's Box will yawn wider: the readouts call for a possible 200 domestic nukes; another 166 are spotted around the globe currently, and the number is expected to increase. There are 119 licensed research reactors nationwide; the Department of Energy (DOE) currently has more than 100 facilities that are ready to be decommissioned, too. The military has more than 145 reactors, most of them placed on nuclear submarines. Specific plans for decommissioning these facilities are hard to come by. The NRC has, however, drawn up some scenarios.

Ironically, mothballing is one of the projections the NRC accepts. It involves first exorcising radioactive fuels and wastes from the structure; later, the reactor is placed in what is referred to in industry parlance as "protective storage." A second scenario has been dubbed "hardened safe storage" or "in-place entombment," depending on the bent of your nuclear mortician.

Both the mothballing and entombment scripts suffer from fatal myopia. Even the Atomic Industrial Forum (AIF), that industry soapbox, admits that these alternatives could require more than 230,000 years of safe storage before such long-lived radionuclides as nickel-59 would decay to acceptable levels.

"Generally, the primary goal of decommissioning should be dismantlement and release of the property on an unrestricted basis at the earliest practical date," writes G.D. Calkins, decommissioning program manager at the NRC. But there are problems with the "decontamination and dismantlement" scenario, too. "D&D" requires costly technology that is largely untested and for which a utility has little economic incentive. Besides this, the community in which the nuclear power plant is located could lose a valuable chunk of its tax base as a result of dismantlement. And because workers employed to cut up the reactor parts and cart them away face severe radiation hazards, the safest alternative for workers may be to place the nuke in "layaway," i.e. mothball it for a period of about 104 years, until short-term radiation problems diminish sufficiently to permit manual dismantlement. But recent discoveries indicate that trace amounts of nickel-59 and niobium-94 may integrate themselves into pressure-vessel steel and would be just as active a century hence, making delayed dismantlement a question mark.

"The lesson of decommissioning is that the whole industry is operating on faith," says George Williams, who watchdogs the "back end" of the nuclear fuel cycle for the California Public Policy Center. Williams knows his nuclear options. The variables in these scenarios are stunning. No one has ever taken apart a major commercial reactor. All the facilities dismantled in the U.S. to date have been small experimental reactors. A 20-megawatt reactor at Santa Susana, California, took two years to tear apart with remotely controlled laser arcs; the task cost $6 million. And a 22-megawatt reactor at Elk River, Minnesota, owned by the Atomic Energy Commission, was dismantled at a cost of 40 percent of its $12 million building outlay (accounting for inflation), after sale to a private utility proved economically unattractive.

Outside of the conceptual realm this is the sum total of Yankee know-how on the decontamination and dismantlement of nuclear power plants, the NRC's "preferred scenario."

By November 1979 PG&E's continued reluctance to reveal its plans inspired the Redwood Alliance to piece together "a first gathering of spirit and expertise toward the decommissioning of Humboldt Bay."

"It's premature to talk about decommissioning. At the appropriate time, we will give detailed plans," was how plant superintendent Edgar Weeks turned down an invite to appear at the gathering and explain the utility's decommissioning position. "They're not serious; they're just having a big party up there," Weeks told local reporters when asked to explain his absence. "The technology exists to decommission a plant. It's not something new and exciting."

Weeks' place on a morning panel designed to inform local residents of the status of the reactor was taken by a cardboard cutout of Reddy Kilowatt, PG&E's long-time logo.

Amory Lovins, the author of *Soft Energy Paths*, set the tone for the Humboldt Bay conference: "PG&E would like to stall decommissioning as long as possible because it's going to be a poor precedent," Lovins told the overflow crowd at Humboldt State University. "The reality is that this plant has to be decommissioned earlier [rather] than later. Although we don't really know how to do that yet, it's a lot safer to dismantle it before it has a chance to slobber more of its contents around the countryside. It's certainly going to be much cheaper if it's done today."

Lovins' message was echoed by Greg Minor, who, fresh out of school back in 1960, had designed the solid-state safety system at Humboldt Bay for General Electric. Minor subsequently made front-page news when he and two other GE engineers quit the megacorp at the height of California's anti-nuke Proposition 15 drive in 1976. Later, he was employed as technical advisor on the movie *The China Syndrome*. His chalk talk on how to dismantle the nuke had the flavor of a visit to a film studio's special effects department.

"Humboldt Bay is an antique," began Minor. "The vessel is smaller and thinner than most recent boiling-water reactors. Also, the reactor's buried underground, which is going to make things difficult.

"First, you're going to have to get the fuel out. You put that underwater for six months before it can be moved off site. Then the primary cooling system is decontaminated by flushing out the pipes with a high-powered acid solvent; this creates a lot of sludge that has to be treated and buried. Next you go after the reactor vessel . . . that's when you bring in the creepy crawlers — high-temperature remote-controlled plasma torches. You do this underwater too, because to cool things off you flood the containment. It's going to be very hot in there and you'll burn out lots of workers getting machinery like remote cranes in places. Robots are one solution but they malfunction too.

"You remove the pressure vessel internals and cut up the vessel. Then it's back to the cooling system and the steam generators with your torches. Next comes the containment building, which is demolished by conventional means — heavy machinery, jackhammers, explosives. Although the surface has been decontaminated, you're still going to get a lot of radioactive dust.

"You go after the auxiliary buildings the same way. The last thing you want to take is the radwaste — you drain and decontaminate the spent-fuel pool, and end up with a lot of 55-gallon drums. Right now you can't ship more than 60,000 curies off site, and Humboldt Bay has at least 16 times that much. Finally, the equipment you've used to decontaminate and dismantle needs to be buried somewhere too. It's all pretty darn fantastic. . . ."

Equally fantastic is the industry's lackadaisical approach to the technological quandaries presented by dismantlement. Writer Hal Rubin, another conference participant, underscored how the utilities are banking on "techno-fixes," the development of tools and techniques as the occasion arises. The "if we can send a man to the moon" clichés are thick on the lips of the regulators too. D.G. Raasch, the head of a Sacramento, California, municipal committee studying the eventual decommissioning of its local nuke, Rancho Seco, was typically blasé when interviewed by Rubin. "If I were trying to do this in five years, I'd be concerned," said Raasch. "For any plant that's 10 to 30 years down the road, I don't have any personal concerns that they won't have the techniques available. We would probably hire someone. By that time, there will probably be some companies that are pretty good at this business."

In testimony before Congress in 1978, however, an AIF spokesperson conceded, "In order to cut up a pressure vessel, you need a large laser arc that we have not developed in this country yet, at least as far as we know."

Although the NRC considers decommissioning to be a health and safety question first, cost-effectiveness has everything to do with a utility's willingness to diligently provide for a decent funeral. The AIF estimates decommissioning expenses would be 10 percent of a plant's construction costs. In the case of Humboldt Bay, that would mean a $2.5 million bill to really close down. But a recent report by one researcher produced an estimate considerably higher than the AIF's.

If all 72 commercial reactors from Maine to California had to be decommissioned today, the costs might be in excess of $4 billion, but no one is knowledgeable enough to provide a "best estimate" on the actual price if one also takes into account all the non-commercial U.S. reactors. The cleanup job at the end of the nuclear fuel cycle would cost many billions of dollars.

What can a handful of part-time, unexpert but zealous freelancers do to prod a giant utility into decommissioning its local nuke? The first thing to be aware of is that a decommissioning battle could give your activist family something to do for several generations. Dismantlement, from permanent closure to final release, could take a few stultifying decades. Decisions come from Washington D.C., and their tedious language often takes the punch out of victories. The decommissioning process, with its built-in delays and turgid legalistics, is definitely not an organization builder.

"It's important that folks bent on decommissioning their local nuke not put all their eggs in lawyers' baskets. Creative confrontation keeps the pot boiling and spurs momentum," says Redwood Alliance office coordinator Carl Zichella. "We haven't yet gone to direct action, but it's a logical way of dramatizing our frustration at the long-windedness of this procedure." Indeed, at the already-closed Humboldt Bay plant, direct action might prove more symbolic than energy-efficient, since neither construction nor operation would be halted by fence crashers.

Nonetheless, the Redwood Alliance does march to the gates ot its local nuke on occasion. It holds poetry readings and on-site meditations, releases balloons, visits the facility in the guise of St. Nuke at Yuletide and petitions the utility to allow representatives on the premises to conduct citizens' seismic investigations. In addition, the Alliance calls weekly press conferences to update the media on progress or lack thereof, conducts phone surveys, keeps "Letters to the Editor" columns buzzing, leaflets the shopping malls and wages hot debate at opportune moments before the county board of supervisors. Such activities are the bulwark of an activist group when a plant is in limbo.

The activists' work also causes management to become considerably disconcerted. In a purloined interoffice memo, Humboldt Bay's Edgar Weeks, after conceding that his company was editing letters from its personnel for publication in the local press, called for a massive P.R. campaign to offset Alliance efforts. "[We have to] develop a good story," he told his superiors.

"The public . . . is only capable of focusing on one nuclear issue at a time. They'll get around to decommissioning," Thomas Cochran, senior staff scientist for the National Resources Defense Council, once said. Toward the end of 1978, even the conservative editors of *Business Week* suggested that "Humboldt Bay may spark the whole decommissioning issue." Indeed, the reactor that fulfilled only 13 years of its potential before it was turned off may prove to be a precursor of the nuclear industry's upcoming 30th birthday — the day when the country's earliest-built reactors will *have* to be decommissioned.

Ironically enough, one of the other leading candidates for decommissioning is also a youngster. Three Mile Island II — a mere fledgling — had only been in operation for several months prior to the celebrated March 1979 "event." Illinois' Dresden I (PG&E helped build it) is 20 years old, but extraordinarily high levels of radiation have curtailed the operators' ability to perform routine maintenance. Decontamination of Dresden's five miles of piping began last spring and will cost $37.5 million — accounting for inflation, about one-third of what it cost to build the nuke in 1959. If flushing out the pipes with a relatively untried Dow chelating agent doesn't work, decommissioning is next on the agenda.

Until recently, officials at Dresden, TMI and Humboldt had been able to treat the end of the fuel cycle as a contingency that might never arise, rather than an inevitability as certain as death. But then, as one wily industry vet tells it, it's always bad business to advertise for the undertakers.

Greenpeace

PART SEVEN

DECOMMISSIONING CALIFORNIA

There is a balancing act occurring in Nuclear California. The same forces that brought the pioneers of atomic energy here also attracted those committed to preserving the environment. Just as the first meltdown of a commercial reactor occurred here, so was the first moratorium passed on nuclear plant construction. Just as the state once had plans for 70 nuclear power plants, so it now bases its forecasts on conservation and renewable energy sources.

This final chapter shows that these changes have occurred because of some tried and true methods used by an aroused citizenry. It is still as Albert Einstein said so long ago: "We must bring the facts about nuclear energy to the village square. From there must come America's answer."

Photo: Greenpeace members block the dumping of nuclear waste off the English coast

HOW TO SECEDE FROM NUCLEAR AMERICA

SAUL BLOOM

So now you've read it. You've found out how the government hid the dangers of atomic testing from thousands of soldiers and the American public; how it has covered up both civilian and military nuclear accidents; how it has bungled for 40 years a solution to radioactive waste storage; and how it even now attempts to bail out a morally and financially bankrupt nuclear industry.

We've tried to show how susceptible the entire nuclear state is to accidents and terrorism; how it needs to spy upon any and all critics of the nuclear establishment; and how even its economic health depends on the manufacture of weapons that fuel the arms race.

California is important not only because of a handful of commercial nuclear plants, but because it is also home to every facet of the nuclear state — to dozens of military and research reactors, hundreds of nuclear bombs, nuclear laundries, plutonium labs and a continuous traffic in high-level nuclear waste. California is important because every one of these activities takes place next to major earthquake faults capable of producing a radioactive disaster unprecedented in American history.

There is a single, common denominator that runs throughout *Nuclear California*: the ionizing radiation that destroys the tissue of living things, a force that makes no distinction between an X-ray or an atomic blast, a routine emission or a reactor core meltdown. The intensities vary, but the fundamental problem is the same. As Singer and Weir wrote in "California's Nuclear Nightmare," "The simple truth is: Nuclear technology is out of control. Any sane governmental policy must be founded on that central fact."

Can we fight back? Of course. Human history is filled with examples of people overcoming apparently insurmountable obstacles. The abolition of slavery, the gaining of women's suffrage, the stopping of the Vietnam War, the eradication of smallpox . . . the message is an old one, but it was never more true — individuals do make a difference.

There are encouraging signs. The growth of nuclear power, particularly in California, has been stopped dead in its tracks. The rapid development of alternative energy sources is making it clear that a non-nuclear future is entirely practical. Hospital workers are starting to question the over-use of medical X-rays. Workers at naval shipyards and nuclear power plants are increasingly blowing the whistle on the shoddy practices which threaten us all. Local citizens and groups have sprung up in California and around the world to challenge the nuclear facilities — both military and civilian — built in their backyards. And a nuclear disarmament movement is finally getting underway in this country, one that will find plenty of support worldwide.

The reporters involved in producing this book were moved by a simple journalistic principle: the search for the truth. Here at Greenpeace, we have decided to take these truths one step farther, and act as responsible citizens in trying to stop this nuclear insanity. We've found that the powerlessness and despair wrought by the Atomic Age are the greatest barriers that stand in our way. But the barriers can be broken down. There are a number of ways which we've found to be effective, ways in which you too can help. Here are some suggestions:

• **Join an Anti-Nuclear Group:** Environmental groups such as Greenpeace and Friends of the Earth have been instrumental in trying to stop the spread of both nuclear power and weapons. Grass-roots organizations such as the Abalone Alliance and the Alliance for Survival also have been in the forefront of opposition to nuclear energy. A small contribution will support their enormously important efforts at lobbying, educating, campaigning, demonstrating and legally intervening. Their publications will help keep you informed with news that all too often doesn't make it in the big media.

• **Pick Your Issue:** Some groups work specifically on the conversion of local nuclear facilities (the Mid-Peninsula Conversion Project and the defense industry). Others are involved in legally challenging the licenses of nuclear plants (Friends of the Earth and Vallecitos, San Onofre). Still others have used civil disobedience against the facilities (Abalone Alliance, Greenpeace and Diablo Canyon). There are also groups which focus exclusively on the arms race (Nuclear Weapons Freeze Campaign).

• **Volunteer:** Anti-nuclear groups are often desperately short of staff. A few hours spent helping out in the office can make a big difference. Professionals can lend such skills as photography, printing and fund-raising.

• **Take Direct Action:** Greenpeace believes that the most effective way to take action is to directly intercede in the events as they occur. For example, in 1970 and 1973 Greenpeace vessels sailed into nuclear testing zones to stop weapons explosions by France and the United States. The group more recently alerted the public to shipments of radioactive waste by placing warning signs along spent-fuel routes throughout the western United States.

Another form of direct action Greenpeace uses is to "bear witness" by simply being present at the event. The group recently used this technique off the coast of England as radioactive wastes were dumped into the ocean. This assures that the public will have a source of information aside from that of the government or the nuclear industry.

- **Form a Local Group:** In your union, professional association, church, college or whatever, help form a group that will work on these issues. Local groups can push for county- or city-wide resolutions on the arms race and nuclear power and lobby for bans on the transport of radioactive waste through the area. Teachers should watch that materials on energy and the arms race are balanced. Hospital workers need to organize against unnecessary exposure. Media workers have a responsibility to educate themselves and present these issues to the public.

Most of the large anti-nuclear groups have local chapters, or you can start your own. We've listed in the bibliography the most prominent ones; contact them for more information. Local actions often make a big difference. People near Ojai in the Santa Barbara Mountains, for example, organized to halt uranium mining by Lomex Corporation. Residents in Pleasanton challenged the expansion of Interstate Nuclear's atomic laundry. And voters in Kern County overwhelmingly rejected an L.A. utility's attempt to site a nuclear plant there.

- **Vote:** Find out the candidates' views on disarmament, nuclear proliferation and atomic power plants. Make sure that nuclear energy is part of the debate. And do vote! Voting can and does make a difference.
- **Write Letters:** It sounds old-fashioned, but politicians *read* their mail and *worry* about it. Richard Nixon worried enough during the Vietnam War that he secretly hired large numbers of fake letter-writers to support his policies.

Above all, get involved. We still have time to stop the dangerous commerce in nuclear waste, to phase out the nuclear facilities in earthquake zones and to create a disarmament movement powerful enough to stop nuclear war. It is a challenge we can't afford to ignore.

Think, for one minute, what the world would be like if there were no nuclear bombs or reactors to worry about.

That's what we're working for.

APPENDIX A

A DIRECTORY OF NUCLEAR CALIFORNIA

I. Major Handlers of Radioactive Materials

Atomics International, 8900 De Soto Ave., Canoga Park, CA 91304; (213) 341-1000.

At its three major plants in the San Fernando Valley AI is licensed by the Nuclear Regulatory Commission to handle 3300 pounds of uranium-235 and more than 17 pounds of plutonium-239. Among its projects are breeder reactor fuel research and reactor decommissioning. AI's Santa Susana Test Facility was the site of this country's first commercial reactor meltdown, in July 1959.

Bechtel Corp., 50 Beale St., San Francisco, CA 94105; (415) 768-1234.

The nation's largest construction and engineering company. Bechtel has built nearly one-half of this country's nuclear power plants and also has the cleanup contract for the damaged Three Mile Island facility.

General Atomic Co., P.O. Box 81608, San Diego, CA 92138; (714) 455-3000.

GA is jointly owned by Scallop Nuclear and Royal Dutch Shell. It is located on a 400-acre site north of San Diego and currently works with the U.S. and Japanese governments to produce a controlled fusion power reactor. GA manufactures the TRIGA research reactor and operates two of these at this site for training and isotope production.

General Electric Co., 175 Curtner Ave., San Jose, CA 95125; (408) 925-1000; and *GE Vallecitos Nuclear Center,* Vallecitos Rd., Sunol, CA 94586; (415) 862-2211.

Currently the world's only builder of the boiling-water class of nuclear reactors. GE has constructed 52 of these plants already and has a $5.5 billion backlog extending into the 1990s. Its nuclear headquarters in San Jose employs 5000 people and serves as a training facility for refueling and maintenance procedures. At its Vallecitos Nuclear Center, near Pleasanton, GE has operated several test reactors and is still licensed by the NRC to handle 150 kilograms of plutonium and large amounts of weapons-grade uranium. Also at Vallecitos is GE's Plutonium Fuels Laboratory, used to design and test reactor components and fuels.

Lawrence Livermore National Laboratory, P.O. Box 808, Livermore, CA 94550; (415) 422-4599.

One of the two designers of nuclear weapons in the United States. Beginning in 1952, LLNL has put its stamp on the development of the neutron bomb, nuclear warheads for the cruise and intercontinental ballistic missiles, and artillery-fired atomic bombshells. The lab is licensed by the Department of Energy to handle at least 495 pounds of weapons-grade uranium and plutonium. It employs 7200 scientists on a 640-acre complex, three miles east of Livermore.

Pacific Gas & Electric, 77 Beale St., San Francisco, CA 94105; (415) 781-4211.

The nation's second largest public utility, providing energy for most of Northern California. PG&E has been a persistent developer of reactors next to major active earthquake faults, including the Vallecitos, Humboldt and Diablo Canyon facilities. After pulling out of Vallecitos, the utility built Humboldt, only to have the NRC close it in 1976. Its nuclear activities in the 1980s have focused on opening its twin reactors at Diablo Canyon. An early pioneer of nuclear power, PG&E has more recently become one of the nation's first utilities to implement far-reaching conservation programs.

Sacramento Municipal Utility District, P.O. Box 15830, Sacramento, CA 95813; (916) 452-3211.

SMUD owns the Rancho Seco nuclear power plant, located 25 miles south of the state capital. The plant is a virtual twin of the damaged Three Mile Island reactor. Since its opening in 1975, Rancho Seco has been plagued by radioactive leaks and shutdowns.

San Diego Gas & Electric, 101 Ash St., San Diego, CA 92101; (714) 232-4252.

Customers served by SDG&E paid the third highest utility rates in the country during 1980. It is co-owner with Southern California Edison of the San Onofre nuclear power plant at San Clemente. Critics have called San Onofre one of the nation's most dangerous nuclear plants due to its numerous technical problems and its proximity to earthquake faults and densely populated areas.

Southern California Edison, 2244 Walnut Grove Ave., Rosemead, CA 91770; (213) 572-1212.

The principal source for electricity in metropolitan Los Angeles. With SDG&E, SoCal Edison has pushed the NRC to license a second reactor at their San Onofre plant, despite serious seismic hazards. By the mid-1980s, Edison hopes to purchase electricity from the Palo Verde reactor in Arizona and bring on line a third reactor at San Onofre.

United States Air Force, The Pentagon, Washington, D.C. 20350; (202) 697-9020.

Defense analysts estimate that the Air Force's Strategic Air Command stores hundreds of nuclear bombs and missiles at three bases in California: Castle Air Force Base near Merced, March Air Force Base near Riverside and Mather Air Force Base near Sacramento. Two air defense bases also reportedly maintain tactical nuclear air-to-air missiles: George Air Force Base, near Barstow in the Mojave Desert, and Castle Air Force Base.

United States Army, The Pentagon, Washington, D.C. 20350; (202) 697-6724; and *United States National Guard,* 2829 Watt Ave., Sacramento, CA 95821; (916) 920-6559.

The Army stores a supply of tactical nuclear weapons at the Sierra Army Depot near Herlong, about 60 miles northwest of Reno, on the Nevada-California border. The National Guard probably maintains tactical nuclear missiles at its air defense unit at the Fresno Air Terminal.

United States Navy, The Pentagon, Washington, D.C. 20350; (202) 697-9020.

The Navy operates as many as 29 nuclear reactors aboard warships in San Diego, San Francisco and Long Beach harbors. Many of these vessels are refueled at the Mare Island Naval Shipyard near San Francisco. Radioactive spills have been reported at both San Diego and Mare Island. The Navy also maintains large stockpiles of nuclear weapons at its arsenals in Concord, Fallbrook, North Island and Seal Beach, and probably at Lemoore and Moffett Field as well.

University of California, Berkeley, CA 94720; (415) 642-6000.

Since 1943, UC has been directly involved in the development, financing and production of nuclear weapons and reactors. Under contract with the Department of Energy, UC manages the Lawrence Livermore National Laboratory (and Los Alamos in New Mexico). Nuclear engineering departments at the Berkeley, Los Angeles and Santa Barbara campuses conduct research for the nuclear industry and training for power plant operators. Each of these campuses operates a small research reactor.

II. Major Defense Contractors

General Dynamics Corp., 5001 Kearny Villa Rd., San Diego, CA 92111; (714) 277-8900.

This St. Louis–based company is the largest defense contractor in the country. California sales totaled $800 million in 1980. At its San Diego facility the company is building the sea-launched Tomahawk cruise missile.

Hughes Aircraft Co., Centinela Ave. and Teale, Culver City, CA 90230; (213) 391-0711.

A subcontractor for the Trident missile system. In 1980, Hughes completed more than $1.5 billion worth of business for the Pentagon, placing it first among California defense contractors.

Lockheed Corp., 2555 North Hollywood Way, Burbank, CA 91520; (213) 847-6121; 1111 Lockheed Way, Sunnyvale, CA 94086; (408) 742-4321.

Lockheed is California's second largest defense contractor, with approximately $1.5 billion in 1980 sales. Company projects for the 1980s include research and development for the Trident missile system.

Rockwell International, 2230 Imperial Highway, El Segundo, CA 90246; (213) 647-5000.

The principal developer of the B-1 bomber. Rockwell is also involved in the propulsion and guidance systems for the MX missile. It ranks fourth among defense contractors in California, with $600 million in 1980 sales.

TRW, Inc., 1 Space Park, Redondo Beach, CA 90278; (213) 535-4321.

TRW handles engineering for the MX missile from its offices at Norton and Vandenberg Air Force bases. Defense contracts for 1980 totaled $400 million in California, placing it sixth among the state's contractors.

III. Major Anti-Nuclear Groups

Abalone Alliance, 2940 Sixteenth St., Room 310, San Francisco, CA 94103; (415) 861-0592; Diablo Project Office, 452 Higuera St., San Luis Obispo, CA 93410; (805) 543-6614.

The statewide network of 65 groups committed to nonviolent direct action and education to stop nuclear power in California. Begun in 1977, the Alliance has focused most of its efforts on the Diablo Canyon plant.

Alliance for Survival, 1473 Echo Park Ave., Los Angeles, CA 90026; (213) 617-2118.

Probably the largest grass-roots anti-nuclear organization in the country. Forty chapters in Southern California are involved in such projects as stopping the San Onofre nuclear plant and the storage of nuclear weapons at the Seal Beach Naval Weapons Station. The Alliance also sponsors the annual "Survival Sunday" at the Hollywood Bowl to raise money for anti-nuclear activities.

American Friends Service Committee, 2160 Lake St., San Francisco, CA 94121; (415) 752-7766.

This long-standing peace group won the Nobel Peace Prize in 1947 for its international work with refugees. In California it has helped initiate the grass-roots anti-nuclear movement. AFSC has conducted nonviolent training for the Abalone Alliance and has been active in promoting alternative energy.

Friends of the Earth, 124 Spear St., San Francisco, CA 94105; (415) 495-4770.

An early opponent of nuclear power, FOE has consistently been a leader in the fight for an alternative energy policy. FOE's legal intervention was instrumental in stopping the construction of nuclear plants at Rancho Seco and Sundesert and helped to revoke the license of GE's test reactor at Vallecitos. They were also intervenors in the Diablo Canyon hearings and more recently have been active in disarmament issues. FOE publishes the monthly newsmagazine *Not Man Apart* and numerous books and pamphlets, and has an estimated 30,000 members nationwide.

Greenpeace, Building E, Fort Mason, San Francisco, CA 94123; (415) 474-6767.

While Greenpeace is best known for its work in preserving marine mammals, during the last 10 years it has repeatedly taken direct action to stop the spread of both nuclear power and nuclear weapons. Greenpeace has confronted nuclear weapons testing in the oceans and in the courts, the dumping of nuclear waste in the oceans and the mining and shipping of uranium ore. More recently, the organization has conducted large-scale poster campaigns along spent-fuel routes to alert the public and push for legislation to ban the shipments. The group is supported by a nationwide membership of 220,000, with about 80,000 in California.

Sierra Club, 530 Bush St., San Francisco, CA 94108; (415) 981-8634.

The largest environmental organization in the country, with more than 245,000 members. The Sierra Club has worked to establish and defend California's nuclear safeguards laws and has been active in legal intervention against the Diablo Canyon and Rancho Seco power plants. It has chapters throughout the state and maintains effective lobbying offices in Sacramento and Washington, D.C.

Natural Resources Defense Council, 25 Kearny St., San Francisco, CA 94108; (415) 421-6561.

NRDC is a national environmental group supported by some 45,000 members. The group does primarily legal work; its San Francisco office has helped defend California's nuclear safeguards laws, worked on nuclear waste disposal and promoted the development of alternative energy in California and the Pacific Northwest. Its two other offices, in New York and Washington, D.C., have been active against the breeder reactor, nuclear proliferation and other developments.

Nuclear Weapons Freeze Campaign, 5480 College Ave., Oakland, CA 94618; (415) 652-5231 / 7250 Franklin Ave., Suite 103, Los Angeles, CA 90046; (213) 850-1683.

A national coalition working to initiate a bilateral halt to new nuclear weapons production and deployment. Some sponsoring organizations, such as the War Resisters League and the Mt. Diablo Peace Center, also are involved in organizing around local nuclear weapons facilities.

University of California Nuclear Weapons Labs Conversion Project, 944 Market St., Room 508, San Francisco, CA 94102; (415) 982-5578.

Begun in 1976 by UC students and faculty and California peace activists to make the university sever its ties with the weapons labs at Lawrence Livermore and Los Alamos. The Labs Project has testified before the regents of the university, conducted studies on conversion possibilities and demonstrated on UC campuses around the state.

IV. Major Pro-Nuclear Groups

California Council for Environmental & Economic Balance, 215 Market St., Suite 930, San Francisco, CA 94105; (415) 495-5666.

A powerful coalition of major corporations and labor unions, largely funded by the energy industry. From its offices in Los Angeles, San Francisco and Sacramento CEEB lobbies and campaigns for the rapid development of nuclear power, weaker air pollution standards and the importation of liquefied natural gas. Members, including CEEB founder ex-Governor Pat Brown, have spearheaded the fight against California's nuclear safeguards laws.

Citizens for Adequate Energy, 333 Kearny St., Suite 707, San Francisco, CA 94108; (415) 392-3210.

The closest thing to a grass-roots, pro-nuclear group in California. CAE is supported by some 4500 individuals and by generous grants from its nearly 100 member corporations. Starting in 1979, CAE has developed educational programs and campaigned for the licensing of Diablo Canyon.

Electric Power Research Institute, 3412 Hillview Ave., Palo Alto, CA 94303; (415) 855-2000.

The major research arm of the nation's electric utilities. EPRI conducts research and development in various energy technologies, with heavy emphasis on nuclear and fossil fuels, and maintains active public relations offices in Palo Alto and Washington, D.C.

V. Regulatory Agencies

California Department of Health Services, 714 P St., Sacramento, CA 95814; (916) 322-2073.

This state agency regulates most of the non-reactor and non-military nuclear material in California, including the licensing and inspection of more than 1800 industrial, medical and research facilities. Through this agency the state is also trying to establish an adequate monitoring system for areas around major military and commercial nuclear facilities.

Local Communities.

Many local governments have passed ordinances banning or restricting the movement of radioactive materials through their jurisdictions. In California, these include the counties of Humboldt, Marin and Sonoma, and the cities of Oakland, Morro Bay and Pismo Beach.

United States Department of Defense, The Pentagon, Washington, D.C. 20350; (202) 545-6700.

The Defense Department is responsible for regulating certain radioactive materials, including nuclear weapons and reactors, at military installations. DOD's Defense Nuclear Agency specifically oversees the military's weapons programs including research into the effects of nuclear weapons, and the government's atomic testing program.

United States Department of Energy, 1333 Broadway, Oakland, CA 94612; (415) 273-7881.

The DOE is responsible for regulating the use of radioactive materials at its major contractors. In California these include the Lawrence Livermore and Lawrence Berkeley laboratories, the Stanford Linear Accelerator and private businesses such as General Atomic and Atomics International. The DOE can, without NRC approval, increase the amount of radioactive material at these facilities. The DOE also designs and maintains the military's arsenal of nuclear weapons and reactors.

United States Nuclear Regulatory Commission, 1450 Maria Lane, Suite 210, Walnut Creek, CA 94546; (415) 943-3700.

The principal federal agency responsible for licensing and inspection of all commercial and most nuclear research facilities in the country, including more than 200 users in California. The NRC's offices are also involved in the research of radiological health effects and waste treatment, and conduct anti-sabotage surveillance of nuclear facilities.

APPENDIX B

A CONSUMER'S GUIDE TO RADIATION

You obviously don't have to be a nuclear physicist to understand the dangers of radiation. We therefore thought it might be useful to provide this guide to radiation and health. We've borrowed here from two excellent sources: "Radiation on the Job" ($2.75 from the Coalition for the Medical Rights of Women, 1638B Haight St., San Francisco, CA 94117) and "Radiation: The Human Cost" (10¢ from the Nuclear Weapons Facilities Project, 514 C St. NE, Washington, D.C. 20002).

The Nuclear Worker

Ask people what a nuclear worker is and they'll probably describe a character sitting in front of the control panel of a nuclear power plant. It's true — that person is indeed a nuclear worker, but the nuclear power industry represents only a fraction of the nuclear workers in this country. Nearly half of the nuclear work force is in the medical industry; these people administer an estimated 700 million X-rays and 70 million nuclear medicine procedures annually. More than 700,000 health workers are exposed to radiation from X-ray and fluoroscopy machines, from activities in nuclear medicine departments and from radioisotopes in hospital laboratories. Furthermore, the nuclear workers in the medical industry must worry not only about radioactive materials, but about radioactive patients as well. Nurses caring for patients with radioactive implants have received average radiation doses of 625 millirems per year, higher than the current occupational dose allowed for non-nuclear workers.

Other nuclear workers are fairly well distributed among the nuclear power industry, government nuclear programs, and education. These people all face serious problems of overexposure because of a history of lax standards for radiation exposure.

A History of Overexposure

The standards for occupational exposure to radiation were set as early as 1900, when it stood at 10 rems *per day*. These limits traditionally have been set up as absolutes — anything above the limit was unsafe, anything below was safe until proven otherwise. Not until the contamination was inflicted and the workers were dead would the limits be lowered. The women in the watchmaking industry during the 1920s, for example, received such a massive exposure to radium that the occupational exposure limit was lowered to 25 rems *per year* in 1938.

This trend has continued during the last four decades. After World War II, the limits were further lowered to 15 rems per year. In 1960 a limit of five rems per year was established, the figure used today for permissible whole body dose to nuclear workers. (Permissible doses to the public, however, are now .025 rems per year.) An increasing body of scientific evidence suggests that even this standard is much too high and that there is in fact no safe limit of exposure to radiation.

Ionizing Radiation

The qualities of radiation apply to a broad category of energy which can be generally broken down into ionizing and non-ionizing types. Forms of non-ionizing radiation include the sun's rays, radio and television signals, visible and invisible light, microwaves and many others. Human beings receive non-ionizing radiation every day without really thinking about it. Some forms such as microwaves, however, may be harmful. Microwave radiation has received publicity recently due to its increased use and controversy over its possible harmful effects. It has been applied widely in health care institutions, in radar and in telephone and TV transmission.

Ionizing radiation is usually referred to in the media as "radiation." It is the result of atoms moving about and colliding and is produced from any nuclear reaction. It is also the reason that nuclear energy poses such a grave health risk. If enough atoms of living tissue are ionized, disease and death of the tissue may result. Some substances, such as plutonium, continue to emit this radiation for thousands of years. There are five types of ionizing radiation, all of them potentially dangerous: alpha and beta particles, gamma rays, X-rays and neutrons.

Alpha particles are relatively large. Although they cannot penetrate our skin, if an alpha emitter is ingested by the body it can do considerable damage to a very small region. Plutonium is an intense alpha emitter.

Beta particles are very small and can be stopped by just a few millimeters of concrete. They can, however, penetrate several centimeters into human tissue. Strontium-90, a common element in nuclear waste, is a beta emitter.

Gamma rays and X-rays are virtually the same except for their origin. They have the same effect on tissue and can travel great distances, easily penetrating the human body. Cobalt-60, another radioactive by-product, emits both gamma and beta radiation.

Neutrons can be dangerous when artificially released from the nucleus of atoms. They can harm body tissue, a principle used in the neutron bomb. Material which absorbs neutrons can become radioactive itself.

Radiation and Health

A number of studies during the last 15 years document the hazards of even small doses of ionizing radiation. Here's a sampling of some of the more significant reports:

Bross Study of Diagnostic X-Rays: A 1978 study led by Dr. Irwin Bross found that people exposed to low-level radiation from medical diagnostic tests may have 10 times as much risk of leukemia as had been established by previously accepted estimates.

Stewart Report on Fetal Radiation Doses: A 1970 study by Dr. Alice Stewart found that a one-rad abdominal dose to pregnant women appears to result in a 50 percent increase in child leukemia before the age of 10. This trend was compared to children of women who received no diagnostic radiation dose to the abdomen during their pregnancy.

Caldwell Study of "Smoky" Veterans: In 1980 Dr. Glyn G. Caldwell completed a study of participants of the August 31, 1957, "Smoky" nuclear test at the Nevada Test Site. Nine cases of leukemia have occurred among the approximately 3200 test participants. The expected incidence of leukemia in this age group is 3.5.

Leukemia cell type	cases observed	cases expected
All types	9	3.5
Acute myelocytic	4	1.1
Chronic myelocytic	3	0.7
Hairy cell	1	—
AML and CML	7	1.8

Johnson Study of Rocky Flats: Dr. Carl Johnson, former director of the Jefferson County Health Department, collected data on the cancer rates of the Coloradans living downwind of Denver's controversial Rocky Flats nuclear weapons plant. In a report released in February 1979, Dr. Johnson found that citizens living in areas contaminated by plutonium were suffering a cancer rate significantly higher than those living in nearby uncontaminated areas.

Cancer Type	% of increase	
	Women	Men
Lung	—	34
Leukemia	—	40
Lymphoma and myeloma	10	43
Colon	30	43
Ovary	24	—
Testis	—	140
Tongue, pharynx, esophagus	100	60

Mancuso Findings at Hanford: In 1964 Dr. Thomas Mancuso was commissioned by the Atomic Energy Commission (forerunner of today's Department of Energy) to look into cancer rates among nuclear workers at the Hanford Nuclear Facility in Washington. He found a rate higher than expected, touching off a major controversy. His funding was terminated by the AEC before completion of the study; analysis of the data continues through independent financial sources. His latest findings are listed below.

Cancer Type	% of increase
All cancers	26
RES neoplasms	58
Bone marrow	107

Wagoner Study of Uranium Miners: As early as 1546 miners of uranium ore were known to suffer from fatal lung diseases. Lung cancer was officially diagnosed among miners in 1879. Despite this long history of documentation, Dr. Joseph Wagoner has continued documenting deaths among uranium miners to convince the United States government that strong enforced regulations are needed to protect all uranium miners.

Lung disease	cases observed*	cases expected*
Nonmalignant respiratory disease	8/80	3.67/24.9
Malignant respiratory disease*	17/144	9.89/29.8

* Indian/white miners

CREDITS

Glenn Barlow has worked for Friends of the Earth and Greenpeace and is a longtime critic of nuclear power in California. He has been directly involved in legal intervention against the Lawrence Livermore, San Onofre and Vallecitos nuclear plants. Barlow's article, "Life Along the Faultlines," was written for *Nuclear California*.

Marc Beyeler is an environmental planner in the Bay Area and has spent the last four years researching nuclear issues for public interest groups. Beyeler's article, "Atomic Assembly Lines," was written for this book.

Saul Bloom is a project coordinator for nuclear issues at Greenpeace's San Francisco office. His article, "How to Secede from Nuclear America," was written for this book.

Marcy Darnovsky is an editor of *It's About Times,* a monthly newspaper of the Abalone Alliance. Her article, "Lawrence Livermore Laboratory," was written for this book.

Congressman Ronald V. Dellums represents California's Eighth Congressional District in the U.S. House of Representatives. He has served in the Congress since 1971 and is currently a member of the Armed Services Committee and chairman of the District of Columbia Committee.

Douglas Foster is a staff writer at the Center for Investigative Reporting. In 1980, he received the first place award for news reporting from the California Newspaper Publishers' Association. His article, "Burial at Sea," is based on an original story entitled "You Are What They Eat," in the July 1981 issue of *Mother Jones* magazine. Foster's other article, "The Meltdown in L.A.," was written for *Nuclear California*.

Jim Harding is Director of Energy Projects for Friends of the Earth. He is a former staff member of the California Energy Commission and is the author of numerous articles on nuclear power and weapons. His story, "The Demise of the Nuclear Industry," was written for this book.

David E. Kaplan is a staff writer at the Center for Investigative Reporting. Portions of his story, "Hottest Roads in California," were excerpted from "A Visit to Oakland's (Radio)Active Port" in the December 9, 1980, issue of *East Bay Express*. "Where the Bombs Are" is condensed from the piece by the same name in the April 1981 issue of *New West* magazine, and is used by permission of the author.

Michael Kepp is a staff member of the Center for Investigative Reporting. His story, "Doin' the Nuclear Wash," is adapted from "Why They Fear the Laundry

Next Door," an article in the March 22, 1981, issue of *California Living,* the Sunday magazine of the *San Francisco Chronicle* and *Examiner.*

Dan Noyes is a staff writer at the Center for Investigative Reporting. His article, "The Emerging Police State," is based on an original story co-authored with David Kaplan entitled, "Treating Nuclear Critics as Enemies," in the August 6, 1981, edition of the *Los Angeles Times.* Portions of "Rehearsals for the Holocaust" are excerpted from "Operation Wigwam," by Dan Noyes, Maureen O'Neill and David Weir in the December 1, 1980, issue of *New West* magazine.

Paul Nussbaum is a reporter in the San Diego bureau of the *Los Angeles Times.* His article, "The Day the Missiles Hit San Diego," originally appeared as "The Day the Missiles Fly... and Hit S.D.," in the March 1, 1981, edition of the *Los Angeles Times.* Copyright © 1981 *Los Angeles Times.* Reprinted by permission.

John Ross is a free-lance journalist in San Francisco. His article, "What to Do with a Dead Nuke," originally appeared as "Taking Apart Your Neighborhood Nuke" in the January 1981 issue of *Mother Jones* magazine, and is reprinted by permission. Copyright © 1981, Foundation for National Progress.

Mark Schapiro is a staff writer at the Center for Investigative Reporting. In 1980, he and David Weir received the National Magazine Award for their work on pesticide dumping. His stories on the MX missile system have been widely syndicated. Schapiro's article, "Atomic Assembly Lines," was written for *Nuclear California.*

Naomi Schiff is an illustrator and designer in Oakland, California. She designed *Nuclear California,* including the cover, and drew all the maps and graphics in the book.

Michael Singer is a news writer at KRON-TV in San Francisco and is an associate of the Center for Investigative Reporting. His story with David Weir, "A City Held Hostage," originally appeared as "Nuclear Nightmare" in the December 3, 1979, issue of *New West* magazine, and is used by permission of the authors. Singer's other piece, "The Silent Epidemic," was written for *Nuclear California.*

Eleanor Smith is a free-lance journalist in San Francisco and was formerly the managing editor of *Not Man Apart,* the monthly magazine of Friends of the Earth. Her article, "Radioactive Research and Development," was written for this book.

Mark Vandervelden is Sacramento bureau chief for California Public Radio. His broadcasts on energy and environmental affairs have been heard on National Public Radio stations throughout the country. Both of his stories, "Controversy at Black Mountain" and "The Radioactive Waste Dilemma," were written for *Nuclear California.*

David Weir is Executive Director of the Center for Investigative Reporting. He is the winner of numerous journalism awards, including the 1980 National Magazine Award for Reporting Excellence. His story with Michael Singer, "A City Held Hostage," originally appeared as "Nuclear Nightmare" in the December 3, 1979, issue of *New West* magazine, and is used by permission of the authors.

The Center for Investigative Reporting is a non-profit organization founded in 1977 to provide in-depth reporting on the individuals and institutions which shape our lives. From its headquarters in California, the staff of the Center has written for *The Nation, New West, Mother Jones, The Los Angeles Times, The Washington Post* and many other publications. The Center also works closely with ABC's "20/20" and CBS's "60 Minutes."

The Center's articles have helped spark Congressional hearings and legislation, U.N. resolutions, public interest lawsuits, and changes in the activities of multinational corporations, government agencies and organized crime figures. Its stories have won the National Magazine Award, the Clarion Award, the National Press Club Award and the Best Censored Story Award.

A combination of commercial fees and the generous support of foundations and private donors allows the Center to be effective despite the many legal and financial constraints on investigative reporting. For more information contact the Center at 1419 Broadway, Rm. 600, Oakland, CA 94612.

Greenpeace is an international organization founded in 1970 to preserve an ecologically clean and safe environment. From its ten regional offices in North America, the Pacific and Western Europe, the group has used non-violent direct action to bring environmental dangers to public attention.

Greenpeace has shut down pirate whaling operations in Taiwan, chased Soviet and Japanese whaling fleets out of the mid-Pacific and off the California coast, and helped to stop nuclear weapons testing in Amchitka and the South Pacific. Greenpeace members also have blocked the transportation and dumping of radioactive waste, shielded newborn harp seals from the blows of hunters, and set dolphins free from fishermen's nets.

In the United States, Greenpeace is supported by 220,000 members and a national network of seven offices. The group's success depends on this membership, strong public support, and the generous contributions that go into its non-profit, direct action campaigns. For membership and further information contact Greenpeace at Bldg. E, Fort Mason, San Francisco, CA 94123.

Nuclear California is a joint project of the Center for Investigative Reporting and Greenpeace. Additional copies of the book can be ordered from Greenpeace/CIR, Bldg. E, Fort Mason, San Francisco, CA 94123. Mail $5.95 plus $1.00 postage and handling. California residents add 35¢ tax.